Histology and Histopathology of the Aging Brain

Interdisciplinary Topics in Gerontology

Vol. 25

Series Editor
H.P. von Hahn, Basel

KARGER

Basel · München · Paris · London · New York · New Delhi · Singapore · Tokyo · Sydney

Histology and Histopathology of the Aging Brain

Volume Editor
Jürg Ulrich, Basel

31 figures and 10 tables, 1988

KARGER

Basel · München · Paris · London · New York · New Delhi · Singapore · Tokyo · Sydney

Interdisciplinary Topics in Gerontology

Library of Congress Cataloging-in-Publication Data
Histology and histopathology of the aging brain / volume editor, Jürg Ulrich.
(Interdisciplinary topics in gerontology; vol. 25)
Includes bibliographies and index.
1. Alzheimer's disease – Pathophysiology. 2. Brain – Aging. 3. Brain – Histopathology.
4. Brain – Histology.
I. Ulrich, Jürg, 1930–. II. Series.
[DNLM: 1. Aging. 2. Alzheimer's Disease. 3. Brain – anatomy and histology.
4. Brain – physiopathology. 5. Dementia, Senile.
W1 IN679 v. 25 / WL 300 H6735] HQ1060.I53 vol. 25
[RC523] 362.6 s-dc 19 [612'.82] DNLM / DLC
ISBN 3-8055-4771-4

Drug Dosage

The authors and the publisher have exerted every effort to ensure that drug selection and dosage set forth in this text are in accord with current recommendations and practice at the time of publication. However, in view of ongoing research, changes in government regulations, and the constant flow of information relating to drug therapy and drug reactions, the reader is urged to check the package insert for each drug for any change in indications and dosage and for added warnings and precautions. This is particularly important when the recommended agent is a new and/or infrequently employed drug.

Printed in Switzerland by Thür AG, Offsetdruck, Pratteln
ISBN 3-8055-4771-4

Contents

IV. Aetiological and Pathogenetic Considerations about Alzheimer Dementia Based on Morphology

Interdiscipl. Topics Geront., vol. 25, pp. VII–IX (Karger, Basel 1988)

Introduction

It is obvious that both brain function and brain morphology change later in life. To the relatives of old persons and to their physicians, increased forgetfulness and sometimes intellectual decay are the most obvious expressions of these alterations. Radiologists might observe that these patients have a considerable cerebral atrophy. Yet in other patients – both with and without dementia – the changes when compared with younger brains might be subtle and detectable only with the microscope. In the first part of the book, these changes are described in groups according to topography. Cortex (chapter 1), hippocampus (chapters 2 and 3) and subcortical nuclei projecting to the cortex (chapter 4) are discussed separately in this part of the book.

A minority of old people do become intensely demented (some 5% of all those reaching an age of more than 65). As our population gets older and older, their number is increasing too and it is estimated that there might be a million of aged in the US and in Europe who need medical and social help because of their dementia. Although the cerebral changes in dementia do not differ in quality from those aged without mental changes, these alterations are present in higher quantity in the demented. The quantitative aspects are treated in chapters 5 and 6.

For those readers who are not involved themselves with histologic work, the most important changes are briefly mentioned here. Their microscopical aspects (with immunohistochemical changes) can be seen in the illustrations for chapters 7 and 11. The changes are (1) intraneuronal tangles described by Alzheimer [1907], figures 1 and 3, chapter 7; (2) the senile plaques [Blocq and Marinesco, 1892], figures 4 and 5, chapter 11; (3) the vascular amyloidosis [Scholz, 1938]; (4) the neuronal granulovacuoles [Simchowitz, 1912] and (5) the Hirano bodies [Hirano, 1966], fig. 1, chapter 7.

Most of the morphological changes observed during normal and pathological aging represent severe alterations of the cytoskeleton, an aspect which is dealt with in the third part. Here the biochemistry of the alterations are analyzed in chapter 8. In chapter 7 these changes are compared with alterations observed in other conditions, which are possibly related to, but different from the Alzheimer changes, such as Pick bodies.

After being introduced into the morphology of these changes the reader will want to be informed on the cause, the etiology of the dementia and its morphological changes. Two of the chapters in part IV discuss possible environmental causes of Alzheimer's disease: chapter 9 viral infections and chapter 10 chronic metallic intoxications. Finally, chapter 11 in part IV, based on very recent investigations, raises the question if Alzheimer's disease consists really of two separate entities: one leading to Alzheimer tangles and one to the plaques. These aspects are discussed in part IV.

During the last two years much progress has been made in the research on the familial form of Alzheimer's disease, including some important aspects of its molecular biology [e.g. Goldgaber et al., 1987; Kang et al., 1987]. As such research is mainly biochemical it is not discussed here, although some authors refer to it.

J. Ulrich, Basel

References

Alzheimer, A.: Über eine eigenartige Erkrankung der Hirnrinde. Zentbl. Neurol. Psychiat. *18:* 177–179 (1907).

Blocq, P.; Marinesco, G.: Sur les lésions et la pathogénie de l'épilepsie dite essentielle. Sem. Méd. *13:* 445–446 (1892).

Divry, P.: Etudes histochimiques des plaques séniles. J. belge Neurol. *27:* 643–657 (1927).

Glenner, G.G.: Alzheimer's disease: multiple cerebral amyloidosis; in Katman, Biological aspects of Alzheimer's disease, Banbury Report 15, pp. 134–144 (Cold Spring Harbor Laboratory, Cold Spring Harbor 1983).

Goldgaber, D.; Lerman, M.I.; McBride, O.W.; Saffiotti, V.; Gajdusek, D.C.: Characterisation and chromosomal localisation of a cDNA encoding brain amyloid of Alzheimer's disease. Science *235:* 877–884 (1987).

Hirano, A.: Neuropathology of amyotrophic lateral sclerosis and Parkinson dementia complex on Guam. Proc. 5th Int. Congr. Neuropathology, Zürich 1965, Int. Congr. Ser. 100, pp. 190–194 (Excerpta Medica, Amsterdam 1966).

Hirano, A.: A guide to neuropathology (Igaku Shoin, Tokyo 1981).

Kahn, J.; Anderton, B.H.; Probst, A.; Ulrich, J.; Esiri, M.M.: Immunohistological study of granulovacuolar degeneration using monoclonal antibodies to neurofilaments. J. Neurol. Neurosurg. Psychiat. *48:* 924–926 (1985).

Kang, J.; Lemaire, H.G.; Unterbeck, A.; Salbaum, J.M.; Masters, C.L.; Grzeschik, K.H.; Multhaup, G.; Beyreuther, K.; Müller-Hill, B.: The precursor of Alzheimer's disease amyloid A 4 protein resembles a cell surface receptor. Nature, Lond. *325:* 733–735 (1987).

Scholz, W.: Studien zur Pathologie der Hirngefässe. II. Die drusige Entartung der Hirnarterien und Kapillaren. Z. ges. Neurol. Psychiat. *162:* 694–715 (1938).

Simchowitz, T.: Histologische Studien über die senile Demenz. Histol. histopath. Arb. Grosshirnrinde *4:* 267–444 (1912).

I. Various Brain Regions during Normal and Pathological Brain Aging

Interdiscipl. Topics Geront., vol. 25, pp. 1–15 (Karger, Basel 1988)

Morphology of the Human Isocortex in Young and Aged Individuals: Qualitative and Quantitative Findings[1]

H. Braak, E. Braak

Zentrum der Morphologie, Johann-Wolfgang-Goethe-Universität,
Frankfurt/Main, BRD

Introduction: Gross Changes

Several structural changes can be demonstrated in the human cerebral cortex as age advances. A few of these changes can be considered an advantage and further differentiation [Curcio et al., 1982; Coleman and Flood, 1987] but most are deleterious to the cortex. The regressive changes, as seen in the senescent brain, mimic, to a certain extent, the characteristics seen in Alzheimer-type dementias [Tomlinson, 1982; Gibson, 1983] and in neuronal ceroid lipofuscinoses [Braak et al., 1984]. The main difference is that the process of age-related cortical destruction progresses slowly and is far less intensive than that seen in the diseased brain.

A gradual decline in brain weight and volume can be recognized from adulthood into old age [Miller et al., 1980; Hubbard and Anderson, 1981a; Haug, 1984]. On gross inspection, narrowing of the gyri and widening of the sulci can be seen, in particular at frontal and temporal poles of the hemispheres. CT scans display a corresponding widening of the subarachnoid space [Earnest et al., 1979]. The decrease in volume of the hemispheres is accompanied by an increased ratio of the gray matter versus the white substance. The volume of the ventricular system depicts an age-associated increase as well [Barron et al., 1976; Earnest et al., 1979; Hubbard and Anderson, 1981b].

[1] Supported by the Deutsche Forschungsgemeinschaft.

Neuronal Constituents of the Isocortex

Most of the neuronal constituents forming the isocortex can either be grouped with the class of pyramidal cells or that of the nonpyramidal cells [H. Braak, 1980, 1984].

Pyramidal Cells and Modified Pyramidal Cells

Pyramidal cells derive their name from the appearance of their cell body. A dominant apical dendrite ascends from a broad stem, traverses suprajacent layers to form a richly arborizing terminal tuft. Thinner dendrites arise from the basal tips of the soma. Both the apical and the basal dendrites generate side branches and all these processes bear a varying number of spiny appendages (fig. 1). The morphology of the dendrites and their endowment with spines vary in characteristic ways as a function of laminar depth and architectural field. Modified pyramidal cells show more or less thoroughgoing variations of their dendritic processes leading to deviations of the typical pyramidal shape of their cell bodies [H. Braak, 1980, 1984; Feldman, 1984].

The axon of pyramidal and modified pyramidal cells arises from the cell body or a dendrite by way of a conical initial segment. It generally runs in a straight course towards the white matter and on its way gives off numerous collaterals. Beyond the initial segment, the axon and some of its collaterals become myelinated. In general, the axon enters the white substance and terminates in other portions of the cortex or subcortical nuclei. Pyramidal cells can, therefore, be considered to function as association, commissural, or projection neurons.

The rounded nucleus of pyramidal and modified pyramidal cells is generally found in a slightly eccentric position. It is pale, of uniform density and contains a conspicuous nucleolus. The soma is endowed with a variable amount of Nissl material that extends into the proximal dendrites. Small lipofuscin granules that can only weakly be stained by aldehyde fuchsin are randomly dispersed throughout wide portions of the soma. The amount of pigment varies from one type of pyramidal cell to another. The axon hillock and the axon itself remain devoid of both Nissl material and lipofuscin granules [H. Braak, 1980, 1984].

Nonpyramidal Cells

Nonpyramidal cells comprise a variety of neuronal types readily distinguishable in Golgi preparations but ill-defined in Nissl-stained material.

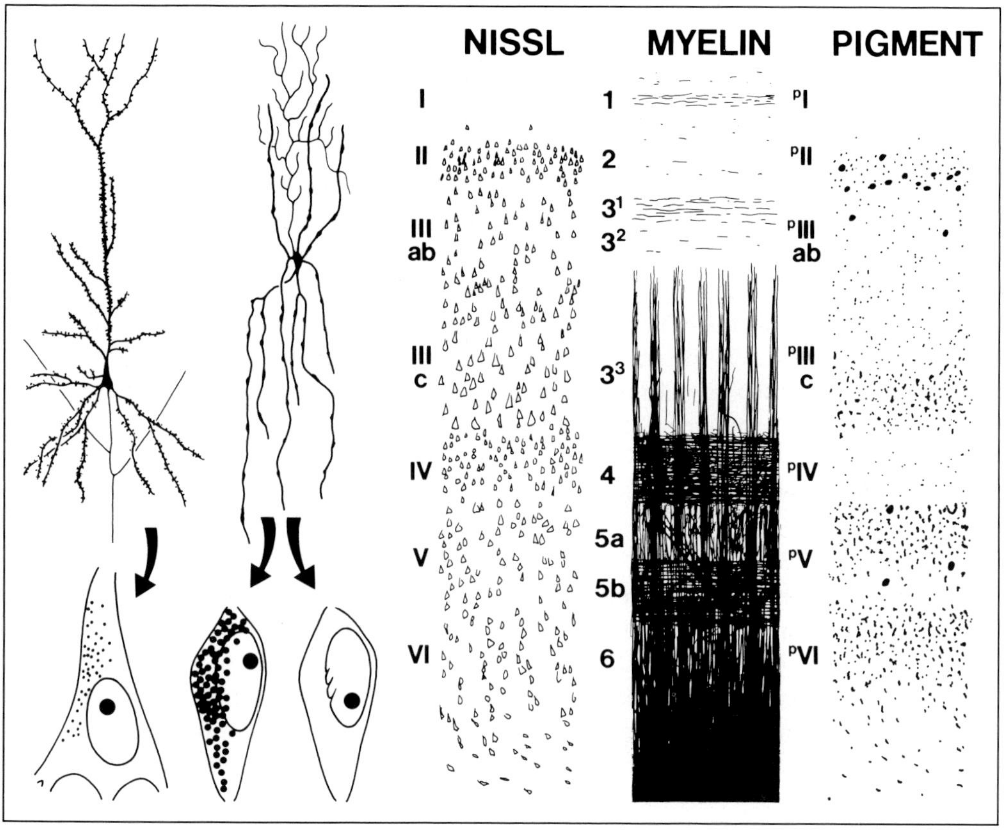

Fig. 1. Left side: Diagram showing the two main classes of cortical nerve cells and their general pattern of pigmentation. The pyramidal neurons generally contain small and faintly stained lipofuscin granules loosely scattered throughout the cell body. The nonpyramidal neurons, in contrast, are either filled with coarse and intensely stained lipofuscin granules or are devoid of pigment. Right side: Synopsis of the fundamental cyto-, myelo-, and pigmentoarchitectonic schemes of the homotypic isocortex.

In general, a typical apical dendrite is lacking. The smoothly contoured and often varicose dendrites emerge abruptly from various points of the soma and branch infrequently. Mature cells show dendrites devoid of spines while immature ones may be endowed with a few of these appendages. The short axon runs in various directions and ramifies into its terminal arborization close to the parent soma [Fairén et al., 1984]. The cells, therefore, can be considered to function as local circuit neurons.

Pigment-Nissl preparations display polygonal or spindle-shaped cell bodies with a pale nucleus that is generally in an eccentric position and shows deep infoldings of the nuclear membrane. Some types of nonpyramidal cells are richly endowed with Nissl material while others show an only sparse amount. The basophilic material does not extend into the dendrites, a feature that helps to differentiate the nonpyramidal cells from pyramidal neurons [E. Braak, 1982]. The pattern of pigmentation differs considerably from that seen in pyramidal cells (fig. 1). Two basic types can be distinguished. One form is filled with coarse and intensely stained lipofuscin granules while the other is devoid of pigment or contains only a few faintly stained granules. Intermediate forms do not exist and, therefore, pigment-Nissl preparations permit differentiation of pyramidal cells from pigment-laden and nonpigmented nonpyramidal cells at light microscopical level (fig. 1).

Pigment-laden nonpyramidal cells are mainly located in supragranular layers II and III and occur in large numbers in phylogenetically advanced portions of the isocortex such as in the opercula of the lateral cerebral sulcus [Schlegelberger and Braak, 1982]. Nonpyramidal cells devoid of pigment prevail in the isocortex and can be encountered in all layers. A large number of them show small elongated and vertically aligned cell bodies meager in Nissl material. At midcortical level, there are voluminous nonpigmented basket cells with a large amount of basophilic material [Sloper et al., 1979; Jones and Hendry, 1984; Braak and Braak, 1986].

Laminar Pattern of the Isocortex

The characteristics described above permit examination of the distribution pattern of the various neuronal constituents in different cortical laminae and areas in the brain of the human adult and they allow evaluation of their ratio to each other.

Evaluation of neuronal densities are fraught with difficulties since the data obtained have to be corrected for tissue shrinkage, secular acceleration, changes in the volume of the neuropil caused by dendritic distortion and increase of glial cells, and several other factors [Haug, 1984]. Evaluation of the ratio of pyramidal cells versus nonpyramidal cells, in contrast, offers the advantage of being independent of all the variables noted above [Braak and Braak, 1986].

The general lamination pattern of the isocortex can be studied in stripes passing through the full depth of the cortex. Brodmann's field 11 is recommended for this purpose since it is formed of a well-differentiated homotypic isocortex with neither a predominance of small or large nerve cells. The boundaries between the layers are easily identified in this frontal association area. The changes in size and packing density of pyramidal cells mainly account for the lamination pattern seen in cell-stained preparations. A synopsis of the fundamental cyto-, myelo-, and pigmentoarchitectonic scheme of the isocortex is depicted in the right half of figure 1.

Quantitative evaluation reveals that, in area 11, the number of pyramidal neurons within the supragranular layers (I–III) roughly matches that of the deep layers (IV–VI). Pyramidal neurons and modified pyramidal cells together form about 85%, and nonpyramidal cells amount to about 15% of all the cortical nerve cells. The nonpyramidal cells are loosely distributed throughout the layers but an abrupt decrease in packing density occurs at the boundary between layers IIIc and IV. Approximately 20% of the nerve cells residing within the supragranular layers (I–III) can be classified as nonpyramidal cells while the deep layers (IV–VI) contain only about 10% of them [Braak and Braak, 1986].

Age-Related Changes of the Isocortex

Neuronal Loss

The severe loss of cortical nerve cells with ageing as described by several authors [Brody, 1955; Brody and Vijayashankar, 1977; Henderson et al., 1975, 1980; Devaney and Johnson, 1980; Anderson et al., 1983] has remained a subject of controversy and recent investigations raise serious doubts as to the extent and even the existence of reduced neuronal densities in the senescent human isocortex [Cragg, 1975; Haug, 1984; Haug et al., 1983, 1984; Braak and Braak, 1986].

Cell Body Changes of Pyramidal Neurons

Some varieties of pyramidal cells display an age-related decrease in the volume of the cell body that gradually assumes an edged and gnarled appearance. This change is most marked in prefrontal layer III pyramidal cells but is hardly recognizable in occipital areas [Haug et al., 1984]. With advancing age, there is a slow and progressive increase of lipofuscin in pigment-containing nerve cells [Brizzee and Ordy, 1981; Rösler and Kemnitz, 1983]. With growing lipofuscin deposits, both the amount of basophilic material within the cytoplasm and the nucleolar volume decline [Mann et al., 1978, 1981].

Dendritic Changes of Pyramidal Neurons

The preservation of the somata of isocortical pyramidal cells into old age does not allow one to assume unimpaired function since the dendritic processes of many of these cells are gradually destroyed (fig. 2). Initial signs of this destructive process are irregular swellings of the pyramidal soma, followed by a progressive loss of dendritic spines and irregular nodulation of the dendrites. The cell gradually becomes denuded from basal dendrites and side branches of the apical dendrite. Finally, the main dendrite is reduced in length and fragmented into filiform portions and irregularly shaped clumps [Scheibel, 1981; Scheibel et al., 1975; Nakamura et al., 1985]. The extent of the destruction varies from one cell to another. It depends largely on the type of the pyramidal cell and its location within a certain layer or area and is prominent in small layer IIIab pyramidal cells of frontal association areas. Even pyramidal cells of the same layer and area show differences in the degree of distortion. As has been pointed out above, the somata of even severely changed pyramidal cells tend to remain in place. The ultimate fate of these cells has not clearly been demonstrated as yet.

Pyramidal cells without signs of deterioration, by contrast, may show particularly elaborate dendritic arbors. There is now growing evidence for a continuous growth of the terminal dendritic segments throughout adulthood into old age, an ability that can be considered a potential for progressive and regenerative changes [Buell and Coleman, 1981; Curcio et al., 1982; Coleman and Flood, 1986].

Fig. 2. Schematic drawings of the morphological changes seen in the aged human isocortex. The dendrites of pyramidal neurons show loss of spines and they are gradually reduced in number and length. Layer IIIab-pyramidal neurons develop spindle-shaped meganeurites filled with lipofuscin granules. Abnormal filaments are found in profiles contributing to the formation of neuritic plaques. The same filaments occur in neurofibrillary tangles and neuropil threads. Loss of myelin is observed in intracortical plexuses. With advancing age, there is a severe numerical reduction of both the pigment-laden and the nonpigmented nonpyramidal neurons.

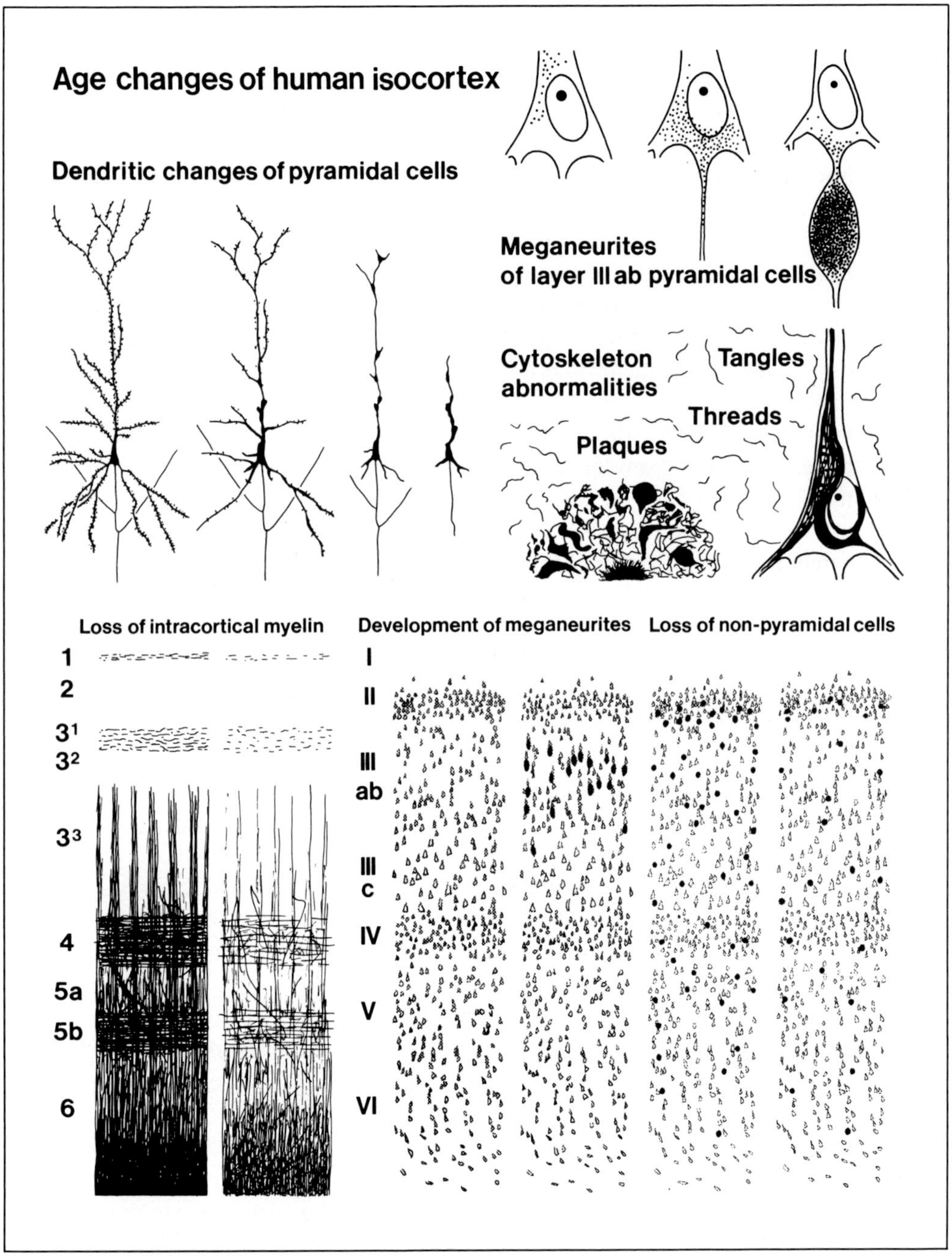
Age changes of human isocortex
Dendritic changes of pyramidal cells
Meganeurites
of layer III ab pyramidal cells
Cytoskeleton
abnormalities
Tangles
Threads
Plaques
Loss of intracortical myelin
Development of meganeurites
Loss of non-pyramidal cells
1
2
3¹
3²
3³
4
5a
5b
6
I
II
III
ab
III
c
IV
V
VI

Development of Meganeurites

Conspicuous changes of the pyramidal cell axon with age have not convincingly been demonstrated as yet with the exception of meganeurite development by layer IIIab pyramidal cells. As has been pointed out above, both the axon hillock and the axon itself normally remain devoid of lipofuscin pigment and Nissl material. The appearance of pigment within the axon is, therefore, a pathological change and this is frequently recognizable in the senile human brain [H. Braak, 1979]. During the first step of this change, the axon hillock is filled with lipofuscin granules. In the next step, perpendiculary arranged rows of a few granules occur in the proximal axon. Further accumulation of pigment finally leads to voluminous spindle-shaped enlargements denoted to as meganeurites (fig. 2) [Purpura and Suzuki, 1976; Braak et al., 1980; Walkley and Pierok, 1986]. The age-related meganeurite development is isocortex-specific and affects only a certain type of pyramidal cell mainly in the layer IIIab. It cannot be recognized in nonpyramidal cells. The change appears to be specific for the ageing human brain. It has in the least not yet been demonstrated in the cerebral cortex of aged nonhuman mammalian species.

Reduction of Myelin

The amount of intracortical myelin is considerably reduced with advancing age. This can most easily be demonstrated in the striate area that normally displays a broad plexus of myelinated fibers referred to as the line of Gennari. This plexus corresponds to the outer line of Baillarger in other isocortical areas. Quantitative measurements of Gennari's line show a gradual and marked myelin reduction with advancing age. The stripes of Baillarger as well as the line of Gennari are considered to be mainly formed of myelinated axon collaterals of suprajacent pyramidal cells and it is probably a change of these pyramidal cells that ultimately accounts for the age-related reduction of intracortical myelin (fig. 2) [Lintl and Braak, 1983].

Loss of subcortical myelin can be encountered as well. The change is most obvious in bundles of cortico-cortical association fibers contributing to the corona radiata [Yakovlev and Lecours, 1967; Kemper, 1984].

Development of Cytoskeleton Abnormalities

Conspicuous changes of the cytoskeleton material can be found in a small number of isocortical neuronal types. This change is a typical concomitant but not an inevitable consequence of advancing age. The pathological fibers present themselves as paired helical filaments with a widest diameter of about 22 nm and constrictions to about 10 nm at every 80-nm interval [Kidd, 1963; Terry, 1963; Wisniewski et al., 1976, 1981]. Basically similar but uncoiled straight filaments may occur as well [Gibson et al., 1976; Yagishita et al., 1981]. The pathological filaments contribute to the formation of three different lesions referred to as neuritic plaques, neurofibrillary tangles, and neuropil threads (fig. 2). The development of these cytoskeleton abnormalities is a unique feature of the human brain. Comparable changes have so far not been recognized in brains of subhuman species even under experimental conditions [Terry and Wisniewski, 1972; Ball et al., 1983].

The globular *neuritic plaques* varying in diameter from 15 to 200 μm are located within the neuropil. They are formed of an intricate feltwork of pathologically changed and often ballooned processes of nerve cells (dendrites as well as axons), reactive astrocytes, microglial cells, and, frequently, a core of extracellularly deposited amyloid. Plaques can be found in all portions of the isocortex. There is, however, a predilection for association areas outside of both motor and sensory core and belt areas. Frequently, the cortex buried in the depth of the sulci is particularly involved. The bulk of the change can generally be found within the supragranular layers II–III. Primary motor and sensory areas and layer IV, by contrast, remain almost spared. Plaques are relatively common in nondemented individuals of old age [Tomlinson et al., 1968; Dayan, 1970; Tomlinson, 1972, 1982; Wisniewski and Terry, 1973; Matsuyama and Nakamura, 1978; Ulrich, 1982; Kemper, 1984; Ulrich and Stähelin, 1984].

Neurofibrillary tangles are formed of tightly packed bundles of paired helical filaments located within the somata of nerve cells whence they may extend into proximal portions of the dendrites. In general, the central bundles of the tangle form a dense and intricate feltwork around the accumulation of lipofuscin granules within the soma. The axon hillock and the proximal axon remain devoid of pathological filaments [Braak and Braak, in press]. Isocortical association areas again bear the bulk of the change, and there is a predilection for layer V and, less expressed, for layer III. Primary motor and sensory areas show the lowest densities of neurofibrillary tangles and layer IV appears to be spared of this change [Jamada and Mehraein, 1968]. Within the isocortex, only pyramidal cells are found to contain tangles [Braak and Braak, 1986] and there is growing evidence that mainly the pyramidal cells furnishing long cortico-cortical connections accumulate the pathological material [Lewis et al., 1986]. In nondemented individuals, isocortical tangle-bearing pyramidal cells are scarce or absent [Tomlinson, 1972, 1982].

Neuropil threads are inconspicuous structures loosely scattered throughout the neuropil. They are formed of small bundles of paired helical filaments contained in slender thread-like profiles that do not cluster or accumulate in patches or columns. Within the isocortex, they are mainly found in areas and layers that show presence of neurofibrillary tangles [Braak et al., 1986]. The axon hillock and the proximal axon remain free of pathological filaments. Transparent Golgi staining combined with demonstration of pathological filaments shows presence of neuropil threads within only the dendrites of pyramidal cells that contain a tangle within their soma. Isocortical nonpyramidal cells and glial cells remain devoid of neuropil threads [Braak and Braak, in press].

Development of Lewy Bodies

Inclusion bodies of the Lewy type show a dense acidophilic core surrounded by a light halo. They are predominantly seen in somata of monoaminergic brain stem neurons but can be recognized in senescent isocortical nerve cells as well, with a predilection for the deep layers V and VI. Lewy bodies are generally considered to be associated with Parkinson's disease, but they can also be encountered in small quantities in elderly individuals without signs of parkinsonism [Woodard, 1962; Forno, 1969]. Until now, Lewy bodies have not been found to occur in brains of subhuman mammalian species.

Changes of Nonpyramidal Cells

Little is known concerning the age-related changes of isocortical nonpyramidal cells. Most of these cells are inconspicuous constituents of the cortex and, therefore, neuronal loss confined to nonpyramidal cells can easily be overlooked. Neither the processes nor the cell body of nonpyramidal cells show marked changes in the senescent brain [Braak and Braak, 1986]. There is, however, growing evidence that many forms of these cells are unusually vulnerable in various pathological conditions [Braak and Goebel, 1978; Sloper et al., 1980; Braak et al., 1983; Braak and Braak, 1986].

Changed Ratio of Pyramidal Cells versus Nonpyramidal Cells

The number and laminar distribution of isocortical pyramidal cells do not show marked changes as age advances. Layer IV, in particular, appears well preserved into old age. The distribution pattern of the nonpyramidal cells as seen in the senile isocortex resembles that found in young adulthood but there is a conspicuous numerical reduction of these neurons (fig. 2). The ratio of pyramidal cells versus nonpyramidal cells changes with age from about 85:15 (preserved up to the sixth decade) to about 92:8 in the ninth decade. Provided that the number of pyramidal cells remains almost constant, this changed ratio indicates a loss of approximately 50% of the nonpyramidal cells. Both the pigment-laden and the nonpigmented forms suffer to about the same extent from this age-associated reduction [Braak and Braak, 1986].

In sum, the total number of isocortical nerve cells appears to be only mildly reduced with age. The cell loss is more or less confined to the class of nonpyramidal cells and these are severely reduced in number. Isocortical pyramidal cells display a rich scale of dystrophic changes but their somata tend to remain in place. The cell bodies and cellular processes of senescent nonpyramidal cells, in contrast, appear almost unchanged, but neurons of this type are, for unknown reasons, particularly prone to dying off with advancing age.

Concluding Remarks

The human isocortex differs in many details from the isocortex of animals. Most of the age-related changes described in the present account manifest themselves only in the human being. The process of ageing, therefore, can barely be understood by extrapolating data from experimental animals. Investigations of the human brain appear, in this light, to be indispensable and irreplaceable for a better understanding of age-related improvements and impairments of human cortical functions.

References

Anderson, J.M.; Hubbard, B.M.; Coghill, G.R.; Slidders, W.: The effect of advanced old age on the neurone content of the cerebral cortex. J. neurol. Sci. *58:* 233–244 (1983).

Ball, M.J.; MacGregor, J.; Fyfe, I.M.; Rapoport, S.I.; London, E.D.: Paucity of morphological changes in the brains of ageing beagle dogs: Further evidence that Alzheimer lesions are unique for primate central nervous system. Neurobiol. Aging *4:* 127–131 (1983).

Barron, S.A.; Jacobs, L.; Kinkel, W.R.: Changes in size of normal lateral ventricles during aging determined by computerized tomography. Neurology *26:* 1011–1013 (1976).

Braak, E.: On the structure of the human striate area; in Beck, Hild, van Limborgh, Ortmann, Pauly, Schiebler, Adv. Anat. Embryol. Cell Biol., vol. 77, pp. 1–87 (Springer, Berlin 1982).

Braak, E.; Braak, H.; Strenge, H.; Muhtaroglu, U.: Age-related alterations of the proximal axon segment in lamina IIIab-pyramidal cells of the human isocortex. A Golgi and fine structural study. J. Hirnforsch. *21:* 531–535 (1980).

Braak, H.: Spindle-shaped appendages of IIIab-pyramids filled with lipofuscin: A striking pathological change of the senescent human isocortex. Acta neuropath. *46:* 197–202 (1979).

Braak, H.: Architectonics of the human telencephalic cortex; in Braitenberg, Barlow, Bizzi, Florey, Grüsser, van der Loos, Studies of brain function, vol. 4, pp. 1–147 (Springer, Berlin 1980).

Braak, H.: Architectonics as seen by lipofuscin stains; in Peters, Jones, Cerebral cortex, vol. 1: Cellular organization of the cerebral cortex, pp. 59–104 (Plenum Press, New York 1984).

Braak, H.; Braak, E.: Golgi preparations as a tool in neuropathology with particular reference to investigations of the human telencephalic cortex. Prog. Neurobiol. *25:* 93–139 (1985).

Braak, H.; Braak, E.: Ratio of pyramidal cells versus non-pyramidal cells in the human frontal isocortex and changes in ratio with ageing and Alzheimer's disease. Prog. Brain Res. *70:* 185–212 (1986).

Braak, H.; Braak, E.: Neuropil threads occur in dendrites of tangle-bearing nerve cells Neuropath. appl. Neurobiol. (in press).

Braak, H.; Braak, E.; Goebel, H.H.: Isocortical pathology in type C Niemann-Pick disease. J. Neuropath. exp. Neurol. *42:* 671–687 (1983).

Braak, H.; Braak, E.; Strenge, H.; Koppang, N.: Canine ceroid lipofuscinosis, a model for ageing of the human isocortex. Gerontology *30:* 215–217 (1984).

Braak, H.; Braak, E.; Grundke-Iqbal, I.; Iqbal, K.: Occurrence of neuropil threads in the senile human brain and in Alzheimer's disease: A third location of paired helical filaments outside of neurofibrillary tangles and neuritic plaques. Neurosci. Lett. *65:* 351–355 (1986).

Braak, H.; Goebel, H.H.: Loss of pigment-laden stellate cells: A severe alteration of the isocortex in juvenile neuronal ceroid-lipofuscinosis. Acta neuropath. *42:* 53–57 (1978).

Brizzee, K.R.; Ordy, J.M.: Age pigments, cell loss and functional implications in the brain; in Sohal, Age pigments, pp. 317–334 (Elsevier, Amsterdam 1981).

Brody, H.: Organization of the cerebral cortex. III. A study of aging in the human cerebral cortex. J. comp. Neurol. *102:* 511–556 (1955).

Brody, H.; Vijayashankar, N.: Cell loss with aging; in Nandy, Sherwin, The aging brain and senile dementia, pp. 15–21 (Plenum Press, New York 1977).

Buell, S.J.; Coleman, P.D.: Quantitative evidence for selective dendritic growth in normal aging but not in senile dementia. Brain Res. *214:* 23–41 (1981).

Coleman, P.D.; Flood, D.G.: Dendritic proliferation in the aging brain as a compensatory repair mechanism. Prog. Brain Res. *70:* 227–237 (1986).

Cragg, B.G.: The density of synapses and neurons in normal, mentally defective and ageing human brains. Brain *98:* 81–90 (1975).

Curcio, C.A.; Buell, S.J.; Coleman, P.D.: Morphology of the aging central nervous system:

Not all downhill; in Mortimer, Pirozzolo, Maletta, Advances in neurogerontology, vol. 3: The aging motor system, pp. 7–35 (Praeger, New York 1982).
Dayan, A.D.: Quantitative histological studies on the aged human brain. 1. Senile plaques and neurofibrillary tangles in 'normal' patients. Acta neuropath. *16:* 85–94 (1970).
Devaney, K.O.; Johnson, H.A.: Neuron loss in the aging visual cortex of man. J. Geront. *35:* 836–841 (1980).
Earnest, M.P.; Heaton, R.K.; Wilkinson, W.E.; Manke, W.F.: Cortical atrophy, ventricular enlargement and intellectual impairment in the aged. Neurology *29:* 1138–1143 (1979).
Fairén, A.; DeFelipe, J.; Regidor, J.: Nonpyramidal neurons: General account; in Peters, Jones, Cerebral cortex, vol. 1: Cellular organization of the cerebral cortex, pp. 201–253 (Plenum Press, New York 1984).
Feldman, M.L.: Morphology of the neocortical pyramidal neuron; in Peters, Jones, Cerebral cortex, vol. 1: Cellular organization of the cerebral cortex, pp. 123–200 (Plenum Press, New York 1984).
Forno, L.S.: Concentric hyalin intraneuronal inclusions of Lewy type in the brains of elderly persons (50 incidental cases): Relationship to parkinsonism. J. Am. geriat. Soc. *17:* 557–575 (1969).
Gibson, P.H.: Form and distribution of senile plaques seen in silver impregnated sections in the brains of intellectually normal elderly people and people with Alzheimer-type dementia. Neuropath. appl. Neurobiol. *9:* 379–389 (1983).
Gibson, P.H.; Stones, M.; Tomlinson, B.E.: Senile changes in the human neocortex and hippocampus compared by the use of the electron and light microscope. J. neurol. Sci. *27:* 389–405 (1976).
Haug, H.: Macroscopic and microscopic morphometry of the human brain and cortex. A survey in the light of new results. Brain Path. *1:* 123–149 (1984).
Haug, H.; Barmwater, U.; Eggers, R.; Fischer, D.; Kühl, S.; Sass, N.L.: Anatomical changes in aging brain. Aging *21:* 1–12 (1983).
Haug, H.; Kühl, S.; Mecke, E.; Sass, N.L.; Wasner, K.: The significance of morphometric procedures in the investigation of age changes in cytoarchitectonic structures of human brain. J. Hirnforsch. *25:* 353–374 (1984).
Henderson, G.; Tomlinson, B.E.; Gibson, P.H.: Counts in human cerebral cortex in normal adults throughout life using an image analysing computer. J. neurol. Sci. *46:* 113–136 (1980).
Henderson, G.; Tomlinson, B.E.; Weightman, D.: Cell counts in the human cerebral cortex using a traditional and an automatic method. J. neurol. Sci. *25:* 129–144 (1975).
Hubbard, B.M.; Anderson, J.M.: A quantitative study of cerebral atrophy in old age and senile dementia. J. neurol. Sci. *50:* 135–145 (1981a).
Hubbard, B.M.; Anderson, J.M.: Age, senile dementia and ventricular enlargement. J. Neurol. Neurosurg. Psychiat. *44:* 631–635 (1981b).
Jamada, M.; Mehraein, P.: Verteilungsmuster der senilen Veränderungen im Gehirn. Die Beteiligung des limbischen Systems bei hirnatrophischen Prozessen des Seniums und bei Morbus Alzheimer. Z. ges. Neurol. Psychiat. *211:* 308–324 (1968).
Jones, E.G.; Hendry, S.H.C.: Basket cells; in Peters, Jones, Cerebral cortex, vol. 1: Cellular organization of the cerebral cortex, pp. 309–336 (Plenum Press, New York 1984).

Kemper, T.: Neuroanatomical and neuropathological changes in normal aging and in dementia; in Albert, Clinical neurology of aging, pp. 9–52 (Oxford University Press, New York 1984).

Kidd, M.: Paired helical filaments in electron microscopy of Alzheimer's disease. Nature, Lond. *197:* 192–193 (1963).

Lewis, D.A.; Campbell, M.J.; Huntley, G.W.; Benson, D.L.; Terry, R.D.; Morrison, J.H.: Laminar and regional specificity of tangle and plaque distribution in Alzheimer's disease (AD): A quantitative study of visual and auditory cortices. Soc. Neurosci. Abstr. *12:* 943 (1986).

Lintl, P.; Braak, H.: Loss of intracortical myelinated fibers: A distinctive age-related alteration in the human striate area. Acta neuropath. *61:* 178–182 (1983).

Mann, D.M.A.; Neary, D.; Yates, P.O.; Lincoln, J.; Snowden, J.S.; Stanworth, P.: Neurofibrillary pathology and protein synthetic capability in nerve cells in Alzheimer's disease. Neuropath. appl. Neurobiol. *7:* 37–47 (1981).

Mann, D.M.A.; Yates, P.O.; Stamp, J.E.: The relationship between lipofuscin pigment and ageing in the human nervous system. J. neurol. Sci. *37:* 83–93 (1978).

Matsuyama, H.; Nakamura, S.: Senile changes in the brain in the Japanese: Incidence of Alzheimer's neurofibrillary change and senile plaques; in Katzman, Terry, Bick, Alzheimer's disease: Senile dementia and related disorders, pp. 287–297 (Raven Press, New York 1978).

Miller, A.K.H.; Alston, R.L.; Corsellis, J.A.N.: Variation with age in the volumes of grey and white matter in the cerebral hemispheres of man: Measurements with an image analyser. Neuropath. appl. Neurobiol. *6:* 119–132 (1980).

Nakamura, S.; Akiguchi, J.; Kameyama, M.; Mizuno, N.: Age-related changes of pyramidal cell basal dendrites in layers III and V of human motor cortex: A quantitative Golgi study. Acta neuropath. *65:* 281–284 (1985).

Pupura, D.P.; Suzuki, K.: Distortion of neuronal geometry and formation of aberrant synapses in neuronal storage diseases. Brain Res. *116:* 1–21 (1976).

Rösler, B.; Kemnitz, P.: Relations between the lipofuscin contents and the numerical density of organelles in the pyramidal cells of the layer-III and layer-V in the area-10 (Brodmann) of frontal lobe of men at different ages. J. Hirnforsch. *24:* 415–424 (1983).

Scheibel, A.B.: The gerohistology of the aging human forebrain: Some structuro-functional considerations; in Enna, Samorajski, Beer, Aging, vol. 17: Brain neurotransmitters and receptors in aging and age-related disorders, pp. 31–42 (Raven Press, New York 1981).

Scheibel, M.E.; Lindsay, R.D.; Tomiyasu, U.; Scheibel, A.B.: Progressive dendritic changes in aging human cortex. Expl Neurol. *47:* 392–403 (1975).

Schlegelberger, T.; Braak, H.: The packing density of supragranular pigment-laden stellate cells in phylogenetically older and newer portions of the human telencephalic cortex. J. Hirnforsch. *23:* 49–54 (1982).

Sloper, J.J.; Hiorns, R.W.; Powell, T.P.S:: A qualitative and quantitative electron microscopic study of the neurons in the primate motor and somatic sensory cortices. Phil. Trans. R. Soc. Lond. (Biol.) *285:* 141–171 (1979).

Sloper, J.J.; Johnson, P.; Powell, T.P.S.: Selective degeneration of interneurons in the motor cortex of infant monkeys following controlled hypoxia: A possible cause of epilepsy. Brain Res. *198:* 204–209 (1980).

Terry, R.D.: The fine structure of neurofibrillary tangles in Alzheimer's disease. J. Neuropath. exp. Neurol. *22:* 629–642 (1963).

Terry, R.D.; Wisniewski, H.M.: Ultrastructure of senile dementia and of experimental analogs; in Gaitz, Aging and the brain, pp. 89–116 (Plenum Press, New York 1972).

Tomlinson, B.E.: Morphological brain changes in non-demented old people; in van Praag, Kalverboer, Ageing of the central nervous system, pp. 38–57 (De Erven F. Bohn N.V. Uitgevers, Haarlem 1972).

Tomlinson, B.E.: Plaques, tangles and Alzheimer's disease. Psychol. Med. *12:* 449–459 (1982).

Tomlinson, B.E.; Blessed, G.; Roth, M.: Observations on the brain of non-demented old people. J. neurol. Sci. *7:* 331–356 (1968).

Ulrich, J.: Senile plaques and neurofibrillary tangles of the Alzheimer type in nondemented individuals at presenile age. Gerontology *28:* 86–90 (1982).

Ulrich, J.; Stähelin, H.B.: The variable topography of Alzheimer type changes in senile dementia and normal old age. Gerontology *30:* 210–214 (1984).

Walkley, S.U.; Pierok, A.L.: Ferric ion-ferrocyanide staining in ganglioside storage disease establishes that meganeurites are of axon hillock origin and distinct from axonal spheroids. Brain Res. *382:* 379–386 (1986).

Wisniewski, H.M.; Narang, H.K.; Terry, R.D.: Neurofibrillary tangles of paired helical filaments. J. neurol. Sci. *27:* 173–181 (1976).

Wisniewski, H.M.; Sinatra, R.S.; Iqbal, K.; Grundke-Iqbal, I.: Neurofibrillary and synaptic pathology in the aged brain; in Johnson, Aging and cell structure, pp. 105–142 (Plenum Press, New York 1981).

Wisniewski, H.M.; Terry, R.D.: Reexamination of the pathogenesis of the senile plaque. Prog. Neuropathol. *2:* 1–26 (1973).

Woodard, J.S.: Concentric hyaline inclusion body formation in mental disease analysis of 27 cases. J. Neuropath. exp. Neurol. *21:* 442–460 (1962).

Yagishita, S.; Itoh, Y.; Wang, N.; Amano, N.: Reappraisal of the fine structure of Alzheimer's neurofibrillary tangles. Acta neuropath. *54:* 239–246 (1981).

Yakovlev, P.J.; Lecours, A.R.: The myelogenetic cycles of regional maturation of the brain; in Minkowski, Regional development of the brain in early life, pp. 3–70 (Blackwell Scientific, Oxford 1967).

H. Braak, MD, Zentrum der Morphologie, Johann-Wolfgang-Goethe-Universität, Theodor-Stern-Kai 7, D–6000 Frankfurt/Main 70 (FRG)

Interdiscipl. Topics Geront., vol. 25, pp. 16–37 (Karger, Basel 1988)

Hippocampal Histopathology – A Critical Substrate for Dementia of the Alzheimer Type

Melvyn J. Ball

Dementia Study Laboratory, University Hospital; Departments of Pathology, Clinical Neurological Sciences and Psychiatry, University of Western Ontario, London, Ont., Canada

'Any article that does not include specific neuropathologic data for each case can hardly be considered a contribution to the study of Alzheimer's disease. (Such investigations are) completely invalidated by the lack of neuropathologic confirmation of AD.'
Meta Neumann
Archs Neurol. *40:* 830, 1983

Introduction

No discussion of the neuropathological changes afflicting the aging human central nervous system would be complete without a critical evaluation of the role of histological abnormalities in the hippocampi, which may account for much of the clinical syndrome known as senile dementia of the Alzheimer type.

The term *hippocampus* was probably first introduced by Arantius in Venice in 1587 [49]. In chapter III of Arantius' book, he referred to an elevation of white substance at the base of the ventricles, which 'faces inward toward the median line ... and is provided with a flexuous figure of varying thickness'. This recalled to Arantius 'the image of a *Hippocampus,* that is, of a little sea-horse ...'.

The first good drawing of the hippocampus in a human brain was provided in 1735 by Duvernoi, a professor of medicine at Tübingen [49], who described the descending horn of the ventricle as ending in a saccular enlargement 'in which we observe a small body of such dazzling whiteness and elegance of structure, that there is scarcely anything handsomer in the whole brain – that is the ... head of the hippocampus of Arantius ... (which) somewhat resembles the figure of a stout silk-worm or that of a ram's horn.'

The Parisian Garengot changed the term in 1742 to *Cornu Ammonis,* or the horn of Ammon, since the Egyptian God Ammon was represented with ram's horns, and the brain representing the organ of the human soul would not have ordinary horns [35].

While such early neuroanatomists knew only the location of this intriguing structure, more detailed anatomical descriptions of the hippocampus appeared considerably later, such as the famous treatise of Ramon y Cajal in 1893, 'On the structure of Ammon's horn' [45].

Many notable texts have been devoted to this fascinating portion of the limbic system, in consideration both of its structure and development [33], and its normal physiological role as a kind of cognitive map [40].

Probably the most striking feature of the clinical syndrome known as Alzheimer's disease, or SDAT (senile dementia Alzheimer type), is the decline in memory function. Given that the hippocampi represent a major anatomical substrate for memory processing, it is surprising that comprehensive reviews of the histopathology of Alzheimer's disease have often put rather limited emphasis upon the cellular abnormalities of this key region. Not one of the 58 chapters in a recently published standard reference text on Alzheimer's dementia [46] was earmarked for this theme. The contribution which follows is dedicated to redressing this deficiency.

Hippocampal Lesions Seen in Alzheimer's Disease

The predilection of the mesial portion of the temporal lobe for the histopathological changes seen both in normal aging of the human nervous system and more prominently in senile dementia of the Alzheimer type has been well documented since the earliest descriptions of these neuronal abnormalities [55, 56]. In a survey of the localization of neurofibrillary degeneration, Hirano and Zimmerman [27] examined the hippocampus, hypothalamus, midbrain, pontine tegmentum and medulla, using silver stains in tissue from 28 mentally normal elderly and 96 neurologically ill patients. These early workers commented that pyramidal nerve cells in two regions – the glomerular formation of the hippocampal gyrus, and the H_1 field of Rose (Sommer sector) in Ammon's horn – were more affected by tangles than all other foci. However, no quantitative data accompanied that impression. In one of the first major symposia devoted to Alzheimer's disease, Corsellis [16] stressed to the other participants how frequently and severely the limbic system was involved in the pathological lesions of Alzheimer's disease and related organic dementias.

The neuronal lesions which many investigators consider to be pathognomonic of Alzheimer's disease include neuritic (senile) plaques, neurofibrillary tangles of Alzheimer, granulovacuolar degeneration of Simchowicz [50], rod-like bodies of Hirano [28], and nerve cell loss or neuronal dropout. Some of these microscopic abnormalities were reported in relationship to the first clinical case described 80 years ago by Alois Alzheimer, the German neurologist after whom the disease has been named [1]. While detailed light microscopic and ultrastructural descriptions of these various lesions can be found in many papers cited in review articles on this disease [7], only a minority of those descriptions have stressed the particular involvement of the mesial temporal lobe. Corsellis [16] had commented that the amygdaloid nucleus, the uncus and the hippocampal region are 'particularly liable to degenerate' in Alzheimer's disease. In 1976, Hooper and Vogel [29] carried out a semiquantitative study in patients with Alzheimer's disease, and surveyed the frontal, temporal and cingulate cortex; the hippocampus and entorhinal cortex; the amygdala; the septal nuclei; the mamillary bodies; and the fornices. When compared to two mentally normal aged control brains, the 9 demented patients' brains studied by these workers showed a 'constant and very severe involvement of the limbic structures, especially the hippocampus'. Hooper and Vogel consequently suggested that this pattern of predilection might even justify the special term, 'limbic dementia'.

Even more curious is the observation that some of these abnormalities are virtually never found *outside* the hippocampal cortex. Ogata et al. [39], in a study of the rod-like eosinophilic structures known as Hirano bodies, surveyed multiple histological sections from various parts of the central nervous system in 240 cases, 67 brains from normal individuals and 173 from pathological cases. In all except one case, Hirano bodies were localized to Ammon's horn, specifically the Sommer sector and the stratum lacunosum beneath it. Similarly, the granulovacuolar degeneration first described by Simchowicz [50] is likewise almost never found anywhere except in the large, pyramidal neurons of the hippocampus [9].

Because this constellation of histopathological changes may also be found in the central nervous system of cognitively normal, elderly people, for a time it was wondered whether Alzheimer's disease merely represented an accelerated form of aging of the human brain. Robert Butler [14] provided an early warning to avoid the conceptual pitfall of simply attributing such changes as 'due to age'. The laboratory application of a type of stereological analysis known as *quantitative morphometry* to postmortem samples of the human brain has definitely laid this misconception to rest. Dayan [18] reported that while the proportion of persons showing neurofibrillary degeneration increases with age in mentally normal subjects, in patients with senile dementia of the Alzheimer type there is a trend towards still larger numbers of neurofibrillary tangles, particularly in Ammon's horn [19]. Tomlinson et al. [55, 56] compared the severity of formation of neuritic plaques, neurofibrillary tangles and granulovacuolar change in 60–70 microscopic fields from the brains of 28 elderly people *without* dementia [55] to these changes in the brains of 50 patients *with* clinically proven dementia [56]. These workers found significantly more severe lesions in the cerebra of those with SDAT. Although granulovacuolar degeneration is similarly increasingly common with normal aging [60], it is also significantly more severe in patients with Alzheimer's dementia [22]. Of special note is Tomlinson's [57, 59] emphasis that by the tenth decade of normal life, virtually every brain will show some neurofibrillary tangles, especially in the hippocampus. In sharp contrast to the hippocampal formation, tangles in the neocortex are either absent or extremely scanty in 80% of old people studied; and neuritic plaques, which are present in 70% of all old persons, are fewer than 10 per low-power microscopic field in 90% of normal subjects [58].

In the Dementia Study Laboratory at the University of Western Ontario, we have used a semiautomated, sampling stage microscope (Wild M501) in quantitative studies of the severity of neurofibrillary tangle formation, granulovacuolar degeneration, and neuronal loss in the serially sectioned hippocampus of normal, aged individuals and of patients considered to have died with Alzheimer's dementia diagnosed by retrospective clinical analysis [2, 3, 12]. These morphometric analyses have demonstrated to my personal satisfaction that the involvement of the hippocampal cortex in Alzheimer's disease is many times more severe than could be attributed to physiological aging – in the case of neurofibrillary tangles, from 6 to 40 times worse; in the case of granulovacuolar degeneration, from 2 to 100 times more severe. Neuronal dropout may be 5 times faster in the hippocampi of people with SDAT than in aged controls [3].

My own emphasis upon hippocampal pathology in Alzheimer's disease has been rooted in the above neuropathological observations, but has also been stimulated by the seminal paper of Wilder Penfield, who with Gordon Mathieson described important findings in two patients who had undergone left temporal lobectomies for intractable epilepsy [41]. Severe, postoperative retrograde amnesia was induced in two men, whose epilepsy originated from one temporal lobe, but whose contralateral hippocampus was postulated as not being functional – a postulate which was subsequently proved at postmortem examination. Penfield hypothesized that, to bring selective portions of our total experiential record into the diencephalon, the hippocampus acts as 'keys of access' to our stream of consciousness. He postulated that 'lines of permanently facilitated connection' are laid down from the hippocampus to many parts of the neocortex throughout an individual's life experience. The retrograde portion of the postoperative amnesia of one of Penfield's patients affected a few months' (preoperative) memories after the first procedure; but when the epilepsy recurred and Penfield later excised an additional more posterior segment of the same hippocampus, the retrograde part of this man's amnesia affected four years of (preoperative) memory stores. Hence Penfield even suggested that these physiological 'lines of facilitation' are laid down beginning at the posterior end of each hippocampal formation in early life, but more and more anteriorly later in life. Because the lesions which we have quantified in this laboratory are not only more severe in the hippocampus of people with Alzheimer's disease than in aged controls, but also show a predilection for the *posterior* half of the hippocampal cortex in SDAT, I have suggested [2] that this posterior involvement may be the anatomical basis for the marked difference between the 'benign forgetfulness of senescence' described by Kral [34], and the devastating incapacitation typical of dementia of the Alzheimer type, in which not only recent but also the remotely learned memories subserving all the activities of daily living have become irretrievable.

Topographical Considerations

Certain astute observers have noticed that, within the hippocampal cortex itself, the cytoarchitectonic distribution of Alzheimer lesions is by no means random, but rather displays a striking and reproducible topographical selectivity. In his clinicopathological analysis of 23 cases, Goodman [23] was probably the first investigator to underscore this important phenomenon. Not only did he state that the pyramidal neurons of the hippocampal cortex were more prone to neurofibrillary degeneration than all other regions, but more specifically that the nerve cells in the 'glomeruli' of the 'substantia reticulata of Arnoldi' exhibited a greater vulnerability to neurofibrillary tangles than did those in any other region of the brain. The first microanatomical 'map' of this regional phenomenon was a line diagram of Ammon's horn and the hippocampal gyrus, in the comprehensive study by Hirano and Zimmerman [27] of the brains of 51 Chamorro patients from Guam (24 with Parkinsonism-dementia complex, 27 with amyotrophic lateral sclerosis) and 103 brains from the Montefiore Hospital, New York (2 with Alzheimer's disease, 3 with Creutzfeldt-Jakob disease, 23 parkinsonians, 39 with ALS, and 36 neurologically and psychiatrically normal elderly control subjects). Their figure 1 showed the distribution of Alzheimer's neurofibrillary degeneration, which was most abundant in the outermost layer of the hippocampal gyrus, the so-called glomerular formation, as well as in Rose's H_1 field, referred to as the Sommer sector. While the amygdaloid nuclei were also involved, they were not so severely afflicted as these two other foci. Hirano and Zimmerman considered that the most vulnerable cells of all were those in the glomerular substance of Arnold (within the entorhinal cortex), which they felt showed earlier and more severe involvement even than those of the H_1-subiculum. They noted, 'There was a rather abrupt increase in these changes at the margin of Sommer's sector.'

These early topographical impressions were eventually substantiated by quantitative studies. In this laboratory, the potentiometric linkage of our scanning stage microscope to an XY pen recorder has permitted the plotting of cytoarchitectonic 'scattergrams' from the sequentially screened hippocampal formations [4, 5]. The density of neurofibrillary tangles, granulovacuolar degeneration and Hirano bodies has been measured in each of six hippocampal 'zones', using a digitizer and programmable calculator (fig. 1). The similarities between the ranking orders for the zones of predilection for these lesions and the 'selective vulnerability' of certain hippocampal zones to hypoxic, ischemic and epileptic effects raised the possibility that some focally accentuated cytotoxic mechanism might be operating in such regional predispositions (see below).

Mounting evidence that such topographical localization was highly germane to a comprehension of the basic pathogenetic mechanisms for organic dementia derived from further evidence of a similar phenomenon in the cytoarchitectonic predisposition of the H_1-subiculum and the entorhinal cortex for the 'inclusion bodies' in cases dying with the unusual

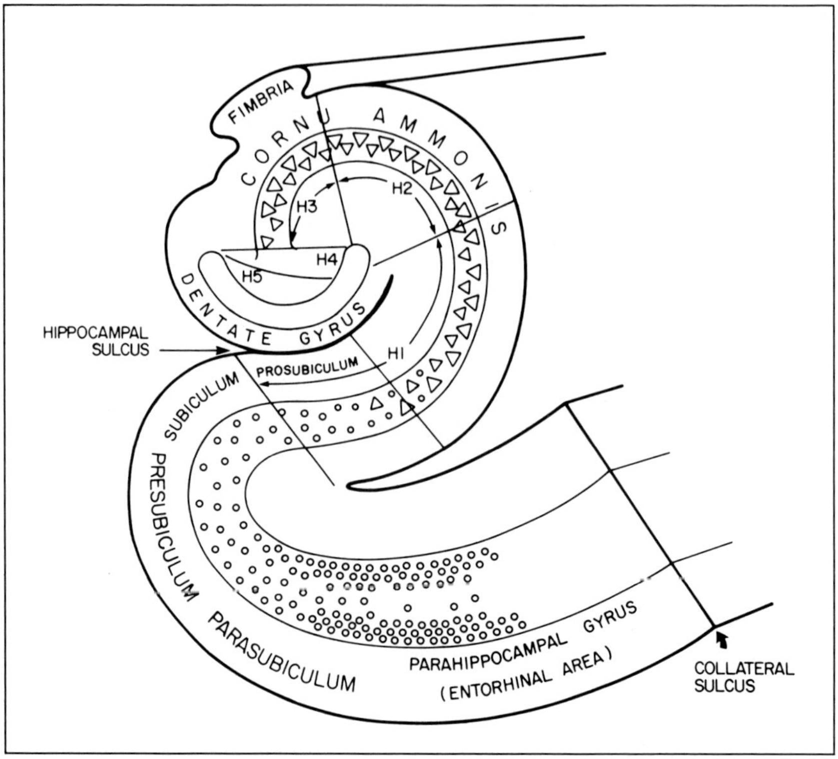

Fig. 1. Microanatomy of the hippocampal formation, showing six cortical regions: endplate (Rose's H_3, H_4 and H_5 fields); the H_2 field; lateral part of H_1 (Sommer sector); medial part of H_1 plus the subiculum; presubiculum, and the entorhinal area.

lobar sclerosis known as *Pick's disease* [6]; and from our additional observations that both the severity of affliction by Alzheimer lesions and their topographical distribution in adults with *Down syndrome* mimic almost exactly these same phenomena in demented patients dying only with Alzheimer's disease [13].

Naturally, the caution advised by Sinex and Myers [51] is well taken when they emphasize, 'Saying that the pathology of Alzheimer's disease is *associated with* dementia is not quite the same as saying that the observable pathology of the disease can *account for* the dementia.' While most neuropathologists agree that virtually every person with Down syndrome beyond the third decade of life will develop plaques and tangles meeting or even exceeding the criteria necessary for a positive diagnosis of Alzheimer's disease [10], the literature suggests that only somewhere between 15 and 30% of people with Down syndrome ever manifest dementia, a clinicopathological discrepancy rightly raised by Wisniewski and Rabe [63].

Despite incomplete associations between the hippocampal pathology and the clinical presentation, one cannot escape the feeling that the recurring commonality of topographical predilection within the hippocampal cortex holds a critical key to the unravelling of the primary pathogenetic mechanism for Alzheimer's disease and related conditions (see below).

Hippocampal Sclerosis Producing an Alzheimer Syndrome: Case Reports

Necropsy findings from the brains of a 63-year-old and a 91-year-old man with a clinical picture compatible with organic dementia of the Alzheimer type may shed additional light upon the question of whether hippocampal dysfunction is the primary event to explain Alzheimer's disease.

At age 57 years, our first patient (L.H.) began to develop a progressive memory deterioration, frequently exhibiting aggressive behavior, mental confusion, and by age 60 the inability to recognize closest friends. There was no pertinent family history. Past history included treatment for bleeding peptic ulcer, cholecystectomy, prostatectomy and gastrectomy. At age 60, a neurologist's examination noted his confusion for time, place and person, and that he did not know his birthdate or the name of the hospital. He had difficulty with simple arithmetic, and was perseverating. He exhibited constructional apraxia, and the neurologist considered his case 'very suggestive for Alzheimer's disease'. The modified Ischemic Score (V. Hachinski) was only 1 out of 18, consistent with a nonvascular or Alzheimer pathology. A psychological test, the Extended Scale for Dementia [25], in which elderly normal subjects score over 230 out of 250, showed values dropping from 138 to as low as 12. A behavioral observation score, the London Psychogeriatric Rating Scale [26], showed increasing disabilities, the worst (highest) possible score on this psychometric instrument being 100. The electroencephalogram was interpreted as normal on two occasions. This man's decline in cognitive function caused him to retire early from his job as a company foreman. Further disorientation and violent behavior eventually led to his admission to the Psychogeriatric Unit, London Psychiatric Hospital (Director, Dr. M. Fisman) at age 61.

Hospital notes recorded that he had been investigated for a scalp lesion, which proved to be herpes zoster; that he had been in some motor vehicle accident many years before, resulting in a possible collapsed vertebra; and that 10 or 20 years before, he had experienced transient numbness of his right arm and transient memory loss, which the patient related to his neck injury and which he felt had been ameliorated by a chiropractor. Hematuria 9 months before death proved to be due to a transitional cell carcinoma of the bladder, and diarrhea 6 months before death, an adenocarcinoma of the rectum. Despite surgical treatment for both lesions, this man died of acute bronchopneumonia at age 63. General autopsy findings included metastatic lesions in liver, bone and regional lymph nodes. The brain weighed only 1,040 g, and the cortex showed a moderate generalized degree of atrophy, especially in the frontal and temporal lobes. The circle of Willis was free of atheroma, and coronal sections of the fixed left cerebrum showed moderate dilatation of the lateral ventricle. The right half of the CNS was frozen at $-70\,^{\circ}C$ for later neurochemical analysis.

Histological analysis of the left cerebrum showed a surprising lack of any neuritic plaques or any neurofibrillary tangles in all neocortical samples, despite a careful survey of ten different regions of the neocortex and a morphometric screening of 4,031 microscopic fields. Moreover, the neuronal population in nearly all of these neocortical regions appeared well preserved. By contrast, however, the cortex of the inferior temporal gyrus, the amygdaloid nucleus, and especially the hippocampal formation showed a devastating degree of nerve cell dropout, accompanied by extremely florid astrocytic gliosis. A very modest number of neurofibrillary tangles, and small numbers of senile plaques, could be found within the hippocampal cortex. A few small veins in the white matter of the inferior temporal lobe showed minimal lymphocytic cuffing. Routine survey of the basal forebrain showed no obvious magnocellular depopulation. Sections of spinal cord were not remarkable.

The nerve cell depletion and accompanying gliosis were sharply delimited, running from the middle portion of the entorhinal cortex and extending into the first part of the adjacent inferior temporal gyrus. Particularly in the absence of any hippocampal granulovacuolar degeneration or rodlike Hirano bodies, and with no neocortical plaques or tangles, this case seems difficult to classify neuropathologically as one of pure or typical Alzheimer's disease. It also does not fit very well with the subacute cortical atrophy originally described by Nevin et al. [38], nor is there any good support for Creutzfeldt-Jakob lesions, for Pick inclusion bodies, or for any other classical organic dementing process. A precise etiology for this man's devastating hippocampal sclerosis remains obscure.

Our second patient (Mr. G.R.) developed a progressive dementing illness at some time during his middle 70s. His activities of daily living became increasingly limited over the next 8 or so years, and by the time he was seen in London, Ontario, he required almost total institutional care and was doubly incontinent. Blood pressure was 140/70, and the only significant clinical diagnosis was senile dementia of the Alzheimer variety. One year before death, he scored only 125 out of 250 on the Extended Scale for Dementia, reflecting a striking decline in his cognitive capabilities, since he had held a job as vice-principal of a secondary school. Six months later he was completely untestable, and following aspiration bronchopneumonia, this man eventually died at age 91 years.

At autopsy the brain, which weighed only 1,075 g, showed mild atrophy of the cortex in the frontal and in the superior temporal regions. The third and lateral ventricles were moderately enlarged. Extensive sampling of ten different neocortical areas from the fixed left cerebrum failed to reveal a single plaque or tangle, although 4,313 microscopic fields were surveyed. Only a slight hint of minimal nerve cell loss was suggested in the middle frontal and the inferior temporal gyri. Despite serial sectioning of the entire left hippocampal formation, not a single neuritic plaque, neurofibrillary tangle, granulovacuolar lesion or Hirano body could be detected. However, a devastating dropout of nerve cells accompanied by severe astrocytic hyperplasia was apparent in the pyramidal layer, especially the H_1 field of Rose and the adjacent subiculum. As with the previous case, the circle of Willis was free of atherosclerosis, and hippocampal branches of the posterior cerebral artery showed no microscopic abnormalities. A few small vessels within the hippocampal white matter were surrounded by a thin cuff of chronic inflammatory cells.

Routine sampling of the nucleus basalis did show a few small foci of probable magnocellular neuronal loss, with old gliosis. As in the previous case, however, no other histological lesions were found anywhere in this CNS to explain this man's progressive dementia. Again this case does not conform to the usual histological criteria for the diagnosis of typical Alzheimer's disease, and once more the precise pathogenetic mechanism underlying the striking hippocampal sclerosis remains uncertain.

In both of these fascinating cases, there are no clues either from the clinical histories or from the neuropathological studies to explain the massive hippocampal destruction. However, neocortical nerve cell abnormalities could not be detected anywhere despite an exhaustive search. Similar findings in the brains of two patients dying in their late 80s have been reported by Ulrich and Stähelin [61]. In our own cases, since rigorous morphometric studies have failed to disclose any unsuspected neocortical lesions, it seems reasonable to conclude that *neuronal damage restricted to the mesial temporal allocortex can produce many,* if not all, *of the clinical features of an organic dementia of the Alzheimer type.* Assays of the choline acetyltransferase enzyme activity from neocortical samples in all lobes of both these patients' brains, kindly performed by Dr. R. Jane Rylett, Department of Physiology, University of Western Ontario, London, have shown that the ChAT activity was within normal limits in both cases, adding further weight to the notion that the hippocampal formation may be the 'locus minoris resistentiae' in an Alzheimer syndrome. Our two cases were histologically and biochemically atypical for what has generally been regarded as 'classical' Alzheimer's disease. There is no doubt, however, that both exceeded the decrement in neuronal population stipulated as one of our morphometric criteria; and that qualified neurologists considered the appropriate diagnosis in both patients to be Alzheimer's disease, clinically.

New Evidence that Hippocampal Dysfunction Alone Accounts for the Dementia of Alzheimer Type

In addition to the morphological and morphometric evidence discussed above, in which the crucial abnormalities of the hippocampal cortex in the syndrome of SDAT cannot be ignored, numerous biochemical investigations of the enzyme activities of the major neurotransmitter markers in Alzheimer's disease have also shown that, of all the areas surveyed, the hippocampus consistently shows the greatest decline in activity [42, 47]. Personal study of the CNS from more than 450 demented individuals has led me to propose that *the decline of all higher cognitive function in Alzheimer's disease can be attributed to histopathological changes in the hippocampal formation, with or without neocortical neuronal lesions* [11].

The dense concentration of motor and sensory fibers in the internal capsule of the brain allows us to explain how a significant deterioration in sensory or motor function, involving a large part of the body, may frequently be attributed not to a widespread destruction of the neocortex, but rather to a restricted lesion, limited to a very small volume of tissue within the internal capsule. If in this sense the capsule represents one type of 'locus minoris resistentiae', the hypothesis promulgated in 1985 is that the hippocampal formation can play a similar role in its physiological concentration of so many memory and higher cortical functions. Variation in the initial clinical presentation of patients with Alzheimer's dementia would merely reflect which microscopic portion of the hippocampal 'funnel' had become dysfunctional first. Another manuscript published by Hyman et al. [31] supports this notion that involvement of the neuronal population of the hippocampal region is a critical anatomical substrate for the dementia of Alzheimer's disease.

In this laboratory, the severity of involvement by neurofibrillary tangles, granulovacuolar change, and nerve cell dropout has already been precisely determined in the serially sectioned hippocampus of 36 patients with Alzheimer's disease, during which a total of 353,353 microscopic fields have been analyzed under the sampling stage microscope. Our morphometric criteria to confirm the category of 'uncomplicated Alzheimer's disease' include: (a) at least 20 tangle-bearing nucleolated neurons/mm^3 (the Adjusted Tangle Index), and/or (b) at least 55 nucleolated neurons/mm^3 showing granulovacuolar degeneration (the Adjusted Granulovacuolar Index), and/or (c) a neuronal population of less than 5,600 nucleolated cells/mm^3 [11]. In every one of these first 36 cases prospectively tested for dementia of the Alzheimer type, this degree of affliction has been found. Although in 15 of these 36, the histological severity of at least one type of lesion falls somewhat short of these morphometric criteria (and of two types in 4 of those 15 cases), in no single instance is the hippocampus not severely affected by at least one of these three classical types of neuronal abnormality. Quantitative morphometry still in progress in this laboratory on several additional cases autopsied since that 1985 publication has likewise failed to turn up any exceptions to these criteria [Ball et al., in preparation]. Of equal interest is the absence of even a single published case in the neuropathological literature to date, in which clinically typical Alzheimer's disease has failed to demonstrate any significant histopathology following *thorough* examination of the hippocampi at postmortem.

Most recently, carefully conducted analyses of clinical and pathological nature, in both demented humans and animal models, have provided impressive support for this concept. Hyman et al. [31, 32] found in 11 cases of Alzheimer's disease that neurofibrillary tangles develop in the perikarya of the neurons of layers II and III of the entorhinal cortex, the cells of origin for the perforant pathway beginning in the parahippocampal gyrus. This path is the principal source of neocortical input to the hippocampal formation. In addition, the termination region for this perforant pathway, in the outer two-thirds of the molecular layer of the dentate gyrus, demonstrates a distinct layer of senile (neuritic) plaques. None of their 8 control subjects showed such lesions. These dramatic alterations effectively *disconnect* the hippocampal formation from the association and limbic cortices. The pyramidal neurons of the H_1-subicular area are also severely affected by neurofibrillary tangles, and since those neurons give rise to the major cortical output from the hippocampus formation, their destruction causes a disconnection of the hippocampal efferents. These workers therefore agree that the pathological changes in the hippocampus have isolated that formation from both cortical afferents and efferents.

Just how widespread the connectivity from the entorhinal region to neocortex may be is dramatically illustrated in a paper by Swanson and Köhler [54], who have employed a sensitive new tracer method using the anterograde axonal transport of a lectin (PHA-L) in the rat, together with a biotin-avidin-HRP immunohistochemical detection system. Injection studies also included double retrograde tracer experiments with fluorescent fast blue and diamidino yellow. They show that the lateral entorhinal area sends fibers to innervate *the entire cortical mantle,* as well as to a longitudinal zone extending for the length of the striatum and the basolateral complex of the amygdala. Although the precise cellular organization of this remarkably extensive association projection is not known, each small PHA-L injection in a different portion of the lateral entorhinal area labels fibers throughout the whole cortical mantle, so that the projection does not appear to be organized in an obviously topographical way. This anatomical study underlines the 'funnel' nature of the mammalian hippocampal formation, which could actually integrate portions of memory engrams from virtually any and all parts of the neocortex.

Eslinger and Damasio [21] have focussed on learning and retention of three types of information in patients with Alzheimer's disease and in normal controls. While 8 patients with the diagnosis of 'probable senile dementia Alzheimer type' (average duration 4 years; in the moderate stage of dementia) were unable to learn series of frequent words and unfamiliar faces, they did improve significantly in a motor skill, showing a learning curve similar to 8 healthy, age-matched controls. Such dissociation in learning has not been previously noted in Alzheimer patients. Their results support the existence of two relatively independent learning systems, related to 'declarative' knowledge and 'procedural' knowledge. The SDAT subjects apparently still have preservation of those neural structures which integrate the visual, motor and kinesthetic information used in such tasks, that permit development and storage of a motor program.

Cohen and Squire [15] have applied these terms, 'procedural' and 'declarative', to the different types of memory. Procedural knowledge refers to motor, perceptual, and cognitive skills which are acquired incrementally and principally by practicing the skill itself. By contrast, declarative knowledge depends on the memory traces of sensory experiences of given stimuli. It ranges in complexity from the memory of a prototype object to complex, autobiographical contextual memories. The striking disparity between procedural and declarative learning in these Alzheimer patients [21] is strong presumptive evidence that SDAT does not randomly or diffusely damage all the anatomical substrates of learning and memory. On the contrary, the cellular damage in Alzheimer's disease must be selective, and should logically involve especially the hippocampal system, which has repeatedly been implicated in the learning and retention of declarative knowledge, but not in the learning of procedural knowledge. Since motor learning might well require different neural connectivities operating independent from the hippocampus, it is noteworthy that most motor system structures appear resistant to the neuropathological ravages found in Alzheimer's disease.

From animal studies involving surgical ablations in primates, it had been suggested that hippocampal damage alone was not sufficient to cause memory dysfunction. Horel [30] had proposed that damage to the 'temporal stem', a white matter fiber system linking the temporal neocortex with subcortical regions, was required to induce medial temporal lobe amnesia. In the same year, an alternative view was promoted by Mishkin [37], who claimed that conjoint damage both to the hippocampus and to the amygdaloid complex is required to produce amnesia. While hippocampal lesions in monkeys would produce statistically significant impairment, combined hippocampal-amygdaloid lesions were considered to produce an even larger memory deficit. Such results from primates left open the question as to whether a human patient with damage limited only to the hippocampal formation would ever demonstrate a substantial memory impairment.

Since the suggestion that amnesia might follow damage to the mesial temporal lobe was first proposed nearly one century ago [62], the recent description of patient R.B. by Zola-Morgan et al. [64] is likely to become as essential to our thinking about this topic as case H.M. of Scoville and Milner [48].

Following coronary artery surgery, R.B., a 52-year-old man, suffered a sudden tear of the atrium, losing 1 liter of blood in less than 2 min, with systolic pressure of 40 mm Hg. During emergency repair of this tear, perhaps 5,000 cm^3 of blood was lost. In the 5 years between that ischemic episode and his death, R.B. was evaluated by extensive neuropsychological tests. He demonstrated little if any retrograde amnesia, but a striking anterograde amnesia, which was apparent not only on tests of free recall but also on tests of recognition memory. Additionally, anecdotal evidence derived from both the patient himself and from people who knew him attested to his anterograde memory impairment. He often would depend on his wife to recount details of recent conversations, and he frequently told the same stories and asked the same questions within a short period. No signs of significant cognitive impairment other than this loss of memory could be demonstrated, nor was there any mention of personality change.

Meticulous neuropathological examination at autopsy, including extremely thorough histological examination, revealed that the only damage which could reasonably be associated with this man's memory defect was a circumscribed destruction of all the neurons in Lorente de No's CA_1 field of the hippocampus, bilaterally. This complete and bilateral neuronal dropout affected the full mediolateral and rostrocaudal extent of this field. There was hyperplasia of glia above and below the empty pyramidal cell layer, but this increased glial density did not extend into the pyramidal layer itself, and was presumed to be a consequence of the abnormally close apposition of cells related to fibers of the alveus and perforant pathway, rather than indicating a reactive process. In the absence of reactive gliosis, the lesions had an old appearance, originating at the time of his initial ischemic event.

The subiculum and CA_3 fields were not involved. A strikingly abrupt transition could be shown between normal neurons in CA_3 and the complete loss of cells in CA_1. Fiber preparations demonstrated that neither the alveus nor the perforant path were affected by the pathology. Fibers from the perforant path travelled through CA_1 to reach their normal termination in the dentate gyrus and CA_3. The fimbria had a normal appearance. No cell loss was detectable in any field of the hippocampal formation except CA_1. The granule cell layer of the dentate gyrus was bilaterally normal, and the molecular layer appeared of normal thickness. All subdivisions of the amygdaloid complex were present and populated by a normal density of nerve cells. There was also no indication of neuronal loss in any components of the mamillary complex, or of the dorsomedial nuclei of the thalamus. There was no gliosis within the fornices or the principal mamillothalamic tracts.

Except for microscopic foci of neuronal loss distributed randomly throughout the neocortex, and representing a very small portion of the total cortical surface, the neocortex was generally normal and had an appropriate laminar organization, with no indication of laminar necrosis. Neither was any cell loss detected in the basal forebrain, with the magnocellular neurons of the nucleus basalis of Meynert present in normal numbers.

The neuropsychological observations in this single case are in accord with results of recent studies in monkeys and rats, in which lesions limited to the hippocampal formation have impaired learning and memory. In particular, experimentally induced ischemia in rats, damaging the CA_1 region of the hippocampus, results in a long-lasting deficit of new learning ability [17]. In animal models, the behavioral deficits are not always quantitatively impressive. Clinical information from case R.B. indicates nevertheless that in man the deficit following this type of hippocampal lesion can be clinically very important. His memory impairment was also stable and long-lasting, since R.B. showed a marked deficit in verbal recall more than 4 years after the injury.

The finding that R.B. was able to recall events occurring prior to his ischemic episode is consistent with the idea that the hippocampus is not an actual *repository* for the storage of remote memories. More likely, it might act as a 'scanner' for the selective assembly of information making up facts and episodes (declarative memory), but not for information underlying skills and procedures (procedural memory).

Zola-Morgan et al. [64] also discuss the neural circuitry of the CA_1 field (Rose's H_1 zone). The hippocampal formation's cytoarchitectonic subdivisions include the dentate gyrus; the hippocampus proper (including CA_1, CA_2 and CA_3); the subicular complex (presubiculum, subiculum and parasubiculum), and the entorhinal cortex. All these fields are linked by a primarily unidirectional flow of information, in which the entorhinal cortex provides the major input, via the perforant pathway, to the dentate gyrus. Although the dentate does not reciprocate this projection, it does provide a major excitatory input to pyramidal neurons of CA_3 via the granule cell axons. Field CA_3 in turn gives rise to the association projections (Schaffer collaterals) to field CA_1, which then projects to the subicular complex, which in turn completes the circuit by projecting to the entorhinal cortex, amongst other regions [64].

Within this circuitry, therefore, a lesion destroying the CA_1 neurons in essence interrupts the chain of information-processing beginning at the dentate and ending in the subicular and entorhinal areas. Such a spatially restricted lesion would nevertheless exert a profound influence on hippocampal function, since the subicular complex and entorhinal cortex are the main efferents from hippocampus to subcortical structures such as the mamillary nuclei and anterior thalamus, and to limbic and cortical structures such as the amygdala, the parahippocampal gyri, and the orbitofrontal cortex. In each cerebrum, the CA_1 field of the human hippocampus alone contains perhaps 4.6 million neurons [64], and in humans, has apparently undergone the greatest enlargement of area relative to the other fields of the hippocampal formation [53].

Of course one must not equate a severe *amnestic* syndrome with the clinical spectrum of incapacitating *dementia* of the Alzheimer type. Nor can one readily explain why patients like Milner's Mr. H.M. have not gone on to develop an Alzheimer dementia, despite careful neuropsychological re-examinations for years after their hippocampal surgical or ischemic ablations. Yet the vast majority of patients with Alzheimer's disease do suffer initially from inability to remember recent events or to learn new lists. Although the analogy with amnesia is imprecise, it can be argued that we should now remove skepticism that discrete lesions of the hippocampal pyramidal cells, whether merely rendered partially dysfunctional (? by neurofibrillary tangles, ? by granulovacuolar degeneration) or totally destroyed (by nerve cell loss), could seriously hamper the 'lines of facilitation' in continuous use for the plethora of physiological phenomena by which memory must very likely underlie most of our higher cognitive capabilities.

Implications for the Etiology of Alzheimer's Disease

A prospective, clinicopathological analysis currently under way in collaboration with Dr. Ronald Doucette (University of Saskatchewan, Saskatoon) has been designed in this laboratory to determine whether the concatenation of microscopic lesions found in patients with Alzheimer's disease in the neocortex, the basal forebrain, or the hippocampal formation best explains the cognitive deterioration. We consider it most likely that the hippocampal pathology shall eventually prove to be the most important. What would such confirmation imply about the basic etiology of senile dementia of the Alzheimer type?

More than a century ago, the selective vulnerability of neurons in the H_1 field of Rose was demonstrated in patients suffering from epilepsy [52]. This discrete region of marked cell loss closely approximates the CA_1 field of Lorente de No, and has come to bear the eponym 'Sommer's sector'.

Depletion of pyramidal neurons in this field of the hippocampus has also been identified as a neuropathological marker of ischemia, both in humans [43] and animal models [44].

It has been proposed that such selective damage during ischemia might be due to the neurotoxic properties of glutamate, aspartate, or some related excitatory amino acid [36]. Zola-Morgan et al. [64] have reviewed the independent observations supporting this idea. There is a high density of glutamate receptors in the CA_1 field of the hippocampus. The major excitatory input to these pyramidal cells (Schaffer collaterals) appears to use glutamate as a transmitter. Injection of an excitatory amino acid antagonist prior to ischemia protects against pyramidal cell loss. And decreasing the synaptic activity of the CA_1 field by eliminating the main excitatory input (the perforant path) also protects the CA_1 cells from ischemic damage.

Could the neurofibrillary degeneration in hippocampal pyramidal neurons which likely precedes their irreversible loss in Alzheimer's disease reflect some similar kind of focal, neurotoxic phenomenon? In this author's 1978 histotopographic study of the cellular changes in Alzheimer hippocampi, this question was considered. 'The similarities between all three orders of predilection and the selective vulnerability of certain hippocampal foci to hypoxic, ischemic and epileptic effects raise the possibility that a *focally accentuated cytotoxic mechanism* might be operating in all such regional predispositions' [4]. The recent induction of paired helical filaments of the Alzheimer type, in human nerve cells in tissue culture using glutamate and aspartate [20], may therefore be of some significance.

A very different speculation, and the theory under current investigation in this laboratory, is that the predilection for hippocampal pyramidal neurons to Alzheimer lesions is based upon the effects of reactivation of the herpes simplex virus, which has been proven to lie in a latent or dormant state in the trigeminal ganglia of man, and might from time to time foray not down those nerve branches to the face or lip (e.g., to cause herpes labialis) but rather inwards along other branches of the trigeminal nerve which innervate the dura and large vessels particularly supplying the medial portion of each temporal lobe [8]. In situ hybridization of postmortem tissue samples, performed through the collaboration of Professor Ashley Haase (University of Minnesota, Minneapolis) has detected viral DNA of the herpes simplex genome in the sensory neurons of the trigeminal ganglion of man [24]. Since the intraperitoneal inoculation of herpes virus in one mouse model leads within a few days to the appearance of herpetic nucleic acid genomic material focally within the rodent hippocampus [Haase, pers. commun. 1986], it is very tempting to speculate that the predilection of the hippocampal formation of man for Alzheimer lesions, and the specific vulnerability of subsets of the hippocampal neurons, might be a microscopic reflection of some neurotropic interaction between the herpes family of viruses and certain, critical neurons of the human brain.

Whatever the eventual explanation for the focality of such lesions in the mesial temporal lobe, those researchers who discover the biological basis for this predilection and focal selectivity will then be within grasp of comprehending the essential etiology for this devastating neurodegenerative disorder.

Acknowledgements

This work was supported by grants from the US National Institutes of Health (2R01 AG 03047 04A1), the Medical Research Council of Canada (PG 21), and Health and Welfare Canada (6606-2889-44). The author thanks Drs. L. Valberg, A.C. Wallace and J. Girvin for encouragement. Drs. N. Jaatoul and V.A. Kral provided clinical diagnoses on the two cases of hippocampal sclerosis. Drs. Jeanette McGlone and John Girvin kindly reviewed, and Mrs. J. Hughes expertly prepared the manuscript. The collaboration of all members of the Dementia Study Group, University of Western Ontario, is gratefully acknowledged, as is the unflagging cooperation of Professors John C.E. Kaufmann and Joseph J. Gilbert.

The scientific standards exemplified in the laboratories of Dr. Anthony Dayan (London, UK) and the late Professor J. Olszewski (Toronto, Canada) provided the initial impetus for these investigations.

References

1 Alzheimer, A.: Über eine eigenartige Erkrankung der Hirnrinde. Allg. Psychiat. *64:* 146–148 (1907).

2 Ball, M.: Neurofibrillary tangles and the pathogenesis of dementia: A quantitative study. Neuropath. appl. Neurobiol. *2:* 395–410 (1976).

3 Ball, M.: Neuronal loss, neurofibrillary tangles and granulovacuolar degeneration in the hippocampus with aging and dementia: A quantitative study. Acta neuropath. *37:* 111–118 (1977).

4 Ball, M.: Histotopography of cellular changes in Alzheimer's disease; in Nandy, Senile dementia: a biomedical approach, pp. 89–104 (Elsevier/North-Holland, New York 1978).

5 Ball, M.: Topographic distribution of neurofibrillary tangles and granulovacuolar degeneration in hippocampal cortex of aging and demented patients: A quantitative study. Acta neuropath. *42:* 73–80 (1978).

6 Ball, M.: Topography of Pick inclusion bodies in hippocampi of demented patients – A quantitative study. J. Neuropath. exp. Neurol. *38:* 614–619 (1979).
7 Ball, M.: Alzheimer's disease – a challenging enigma. Archs Pathol. Lab. Med. *106:* 157–162 (1982).
8 Ball, M.: Limbic predilection in Alzheimer dementia: Is reactivated herpes virus involved? Can. J. neurol. Sci. *9:* 303–306 (1982).
9 Ball, M.: Granulovacuolar degeneration, chap. 7; in Reisberg, Alzheimer's disease – The standard reference, pp. 62–68 (Free Press-Macmillan, New York 1983).
10 Ball, M.: Pathological similarities between Alzheimer disease and Down syndrome: Is there a genetic link? Integr. Psychiat. *5:* 159–170 (1987).
11 Ball, M.; Hachinski, V.; Fox, A.; Kirshen, A.; Fisman, M.; Blume, W.; Kral, V.; Fox, H.; Merskey, H.: A new definition of Alzheimer's disease: A hippocampal dementia. Lancet *i:* 14–16 (1985).
12 Ball, M.; Lo, P.: Granulovacuolar degeneration in the aging brain and in dementia. J. Neuropath. exp. Neurol. *36:* 474–487 (1977).
13 Ball, M.; Schapiro, M.; Rapoport, S.: Neuropathological relationships between Down syndrome and senile dementia Alzheimer type; in Epstein, The neurobiology of Down syndrome, pp. 45–58 (Raven Press, New York 1986).
14 Butler, R.: Alzheimer's disease – senile dementia and related disorders: The role of NIA; in Katzman, Terry, Bick, Alzheimer's disease: Senile dementia and related disorders, pp. 5–9 (Raven Press, New York 1978).
15 Cohen, N.; Squire, L.: Preserved learning and retention of pattern-analysing skill in amnesia: Dissociation of knowing how and knowing that. Science *210:* 207–210 (1980).
16 Corsellis, J.: The limbic areas in Alzheimer's disease and in other conditions associated with dementia; in Wolstenholme, O'Connor, Alzheimer's disease. Ciba Fdn Symp., pp. 37–50 (Churchill, London 1970).
17 Davis, H.; Tribuna, J.; Pulsinelli, W.; Volpe, B.: Reference and working memory of rats following hippocampal damage induced by transient forebrain ischemia. Physiol. Behav. *37:* 387–392 (1986).
18 Dayan, A.: Quantitative histological studies on the aged human brain. I. Senile plaques and neurofibrillary tangles in normal patients. Acta neuropath. *16:* 85–94 (1970).
19 Dayan, A.: Quantitative histological studies on the aged human brain. II. Senile plaques and neurofibrillary tangles in senile dementia. Acta neuropath. *16:* 95–102 (1970).
20 Boni, U. de; Crapper-McLachlan, D.: Controlled induction of paired helical filaments of the Alzheimer type in cultured neurons by glutamate and aspartate. J. neurol. Sci. *68:* 105–118 (1985).
21 Eslinger, P.; Damasio, A.: Preserved motor learning in Alzheimer's disease: Implications for anatomy and behavior. J. Neurosci. *6:* 3006–3009 (1986).
22 Gibson, P.; Tomlinson, B.: Numbers of Hirano bodies in the hippocampus of normal and demented people with Alzheimer's disease. J. neurol. Sci. *33:* 199–206 (1977).
23 Goodman, L.: Alzheimer's disease, a clinico-pathologic analysis of twenty-three cases with a theory on pathogenesis. J. nerv. ment. Dis. *118:* 97–130 (1953).
24 Haase, A.; Lewis, E.; Wietgrefe, S.; Zupancic, M.; Diedrich, J.; Minnigan, H.; Ball,

M.: Speculations on the role of transmissible agents in the pathogenesis of Alzheimer's disease. Can. J. neurol. Sci. *13:* 449–451 (1986).
25 Hersch, E.: Development and application of the extended scale for dementia. J. Am. geriat. Soc. *27:* 348–354 (1979).
26 Hersch, E.; Kral, V.; Palmer, R.: Clinical value of the London Psychogeriatric Rating Scale. J. Am. geriat. Soc. *26:* 348–354 (1978).
27 Hirano, A.; Zimmerman, H.: Alzheimer's neurofibrillary changes: A topographic study. Archs Neurol. *7:* 227–242 (1962).
28 Hirano, A.; Dembitzer, H.; Kurland, L.; Zimmerman, H.: The fine structure of some intraganglionic alterations: Neurofibrillary tangles, granulovacuolar bodies and 'rod-like' structures as seen in Guam amyotrophic lateral sclerosis and parkinsonism-dementia complex. J. Neuropath. exp. Neurol. *27:* 167–182 (1968).
29 Hooper, M.; Vogel, F.: The limbic system in Alzheimer's disease. Am. J. Path. *85:* 1–19 (1976).
30 Horel, J.: The neuroanatomy of amnesia: A critique of the hippocampal memory hypothesis. Brain *101:* 403–445 (1978).
31 Hyman, B.; Van Hoesen, G.; Damasio, A.; Barnes, C.: Alzheimer's disease: cell-specific pathology isolates the hippocampal formation. Science *225:* 1168–1170 (1984).
32 Hyman, B.; Van Hoesen, G.; Kromer, L.; Damasio, A.: Perforant pathway changes and the memory impairment of Alzheimer's disease. Ann. Neurol. *20:* 472–481 (1986).
33 Isaacson, R.; Pribram, K.: The hippocampus, vol. 1: Structure and development (Plenum Press, New York 1975).
34 Kral, V.: Benign senescent forgetfulness; in Katzman, Terry, Bick, Alzheimer's disease: Senile dementia and related disorders. Aging, vol. 7, pp. 47–51 (Raven Press, New York 1978).
35 Lewis, F.: The significance of the term hippocampus. J. comp. Neurol. *35:* 213–230 (1923).
36 Meldrum, B.: Excitatory amino acids and anoxic/ischemic brain damage. Trends Neurosci. *1985:* 47–48 (Feb. 1985).
37 Mishkin, M.: Memory in monkeys severely impaired by combined but not by separate removal of amygdala and hippocampus. Nature, Lond. *273:* 297–298 (1978).
38 Nevin, S.; McMenemy, W.; Behrman, S.; Jones, D.: Subacute spongiform encephalopathy – a subacute form of encephalopathy attributable to vascular dysfunction (spongiform cerebral atrophy). Brain *83:* 519–564 (1960).
39 Ogata, J.; Budzilovich, G.; Cravioto, H.: A study of rod-like structures (Hirano bodies) in 240 normal and pathological brains. Acta neuropath. *21:* 61–67 (1972).
40 O'Keefe, J.; Nadel, L.: The hippocampus as a cognitive map (Claredon Press, Oxford 1978).
41 Penfield, W.; Mathieson, G.: Autopsy findings and comments on the role of hippocampus in experiential recall. Archs Neurol. *31:* 145–154 (1974).
42 Perry, E.; Perry, R.: The cholinergic system in Alzheimer's disease; in Roberts, Biochemistry of dementia, pp. 135–183 (Wiley, New York 1980).
43 Plum, F.: What causes infarction in ischemic brain? Neurology *33:* 222–233 (1983).
44 Pulsinelli, W.; Brierley, J.: A new model of bilateral hemispheric ischemia in the unanaesthetised rat. Stroke *10:* 267–272 (1979).

45 Ramon y Cajal, S.: The structure of Ammon's horn (Thomas, Springfield 1968).
46 Reisberg, B.: Alzheimer's disease – the standard reference (Free Press-Macmillan, New York 1983).
47 Rossor, M.; Garrett, N.; Johnson, A.; Mountjoy, C.; Roth, M.; Iverson, L.: A postmortem study of the cholinergic and GABA systems in senile dementia. Brain *105:* 313–330 (1982).
48 Scoville, W.; Milner, B.: Loss of recent memory after bilateral hippocampal lesions. J. Neurol. Neurosurg. Psychiat. *20:* 11–21 (1957).
49 Seifert, W.: Hippocampus symposium panel discussion, part IV, Epilogue; in Seifert, Neurobiology of the hippocampus, pp. 624–627 (Academic Press, New York 1983).
50 Simchowicz, T.: Histopathologische Studien über die senile Demenz; in Nissl, Alzheimer, Histologie und histopathologische Arbeiten über die Grosshirnrinde, vol. 4, pp. 267–444 (Fischer, Jena 1911).
51 Sinex, F.; Myers, R.: Alzheimer's disease, Down's syndrome, and aging: the genetic approach. Ann. N.Y. Acad. Sci. *396:* 3–13 (1982).
52 Sommer, W.: Erkrankung des Ammonshorns als aetiologisches Moment der Epilepsie. Arch. Psychiat. NervKrankh. *10:* 631–675 (1880).
53 Stephan, H.: Evolutionary trends in limbic structures. Neurosci. Biobehav. Rev. *7:* 367–374 (1983).
54 Swanson, L.; Köhler, C.: Anatomical evidence for direct projections from the entorhinal area to the entire cortical mantle in the rat. J. Neurosci. *6:* 3010–3023 (1986).
55 Tomlinson, B.; Blessed, G.; Roth, M.: Observations on the brains of non-demented old people. J. neurol. Sci. *7:* 331–356 (1968).
56 Tomlinson, B.; Blessed, G.; Roth, M.: Observations on the brains of demented old people. J. neurol. Sci. *11:* 205–242 (1970).
57 Tomlinson, B.: Morphological changes and dementia in old age; in Smith, Kinsbourne, Aging and dementia, pp. 25–26 (Prentice-Hall, New York 1977).
58 Tomlinson, B.: The pathology of dementia; in Wells, Dementia; 2nd ed., pp. 113–153 (Davis, Philadelphia 1977).
59 Tomlinson, B.: The aging brain; in Smith, Cavanagh, Recent advances in neuropathology, pp. 129–159 (Churchill Livingstone, New York 1979).
60 Tomlinson, B.; Kitchener, D.: Granulovacuolar degeneration of hippocampal pyramidal cells. J. Path. *106:* 165–185 (1972).
61 Ulrich, J.; Stähelin, H.: The variable topography of Alzheimer type changes in senile dementia and normal old age. Gerontology *30:* 210–214 (1984).
62 Von Bechterew, W.: Demonstration eines Gehirns mit Zerstörung der vorderen und inneren Teile der Hirnrinde beider Schläfenlappen. Neurol. Zentbl. *19:* 990–991 (1900).
63 Wisniewski, H.; Rabe, A.: Discrepancy between Alzheimer-type neuropathology and dementia in people with Down syndrome. Ann. N.Y. Acad. Sci. (in press, 1987).
64 Zola-Morgan, S.; Squire, L.; Amaral, D.: Human amnesia and the medial temporal region: Enduring memory impairment following a bilateral lesion limited to field CA_1 of the hippocampus. J. Neurosci. *6:* 2950–2967 (1986).

Melvyn J. Ball, MD, Dementia Study Laboratory, University Hospital;
Departments of Pathology, Clinical Neurological Sciences and Psychiatry,
University of Western Ontario, London, Ont. N6A 5C1 (Canada)

Interdiscipl. Topics Geront., vol. 25, pp. 38–58 (Karger, Basel 1988)

Somatostatin and Neuropeptide Y: Alterations and Coexistence in the Alzheimer's Dementia Hippocampus[1]

Victoria Chan-Palay

Neurology Clinic, University Hospital, Zürich, Switzerland

Introduction

Alzheimer's type dementia (ATD) is characterized clinically by dementia and histologically by the presence of numerous neuritic or senile plaques and neurofibrillary tangles in the neocortex and hippocampus [Tomlinson et al., 1970; Tomlinson and Corsellis, 1984]. The degree of dementia is reported to be positively correlated with the number of cortical plaques [Blessed et al., 1968; Perry et al., 1978] and cortical and hippocampal neurofibrillary tangles [Wilcock and Esiri, 1982].

Of the numerous neurotransmitter systems studied in postmortem brains of patients with ATD, reduction of choline acetyltransferase (CAT) activity has been most consistently found [Bowen et al., 1976; Davies and Maloney, 1976; Perry et al., 1977; Rossor et al., 1982]. This decrease of cortical and hippocampal cholinergic activity has been attributed to a loss of cholinergic innervation from the basal forebrain, the septal nuclei and the nucleus of Meynert [Whitehouse et al., 1982; Coyle et al., 1983; Rossor et al., 1982; Henke and Lang, 1983; McGeer et al., 1984]. In addition to these well-documented cholinergic abnormalities, noradrenergic and serotonin deficits [Hardy et al., 1985; Gottfries et al., 1983; Carlsson et al., 1980; Cross et al., 1983; Berger and Gaspar, this volume] have also been reported, indicating degeneration of projections from the locus coeruleus and the raphe nuclei. More recently, reductions of somatostatin concentrations have been demonstrated in the cerebral cortex [Hardy et al., 1985;

[1] Supported in part by United States AFOSR grant 86-0176.

Davies et al., 1980; Ferrier et al., 1983; Davies and Terry, 1981; Rossor et al., 1982] and somatostatin-immunoreactive axons have been shown to participate in the formation of neuritic plaques [Chan-Palay, 1987; Armstrong et al., 1985; Morrison et al., 1985; Nakamura and Vincent, 1986] and neurofibrillary tangles [Roberts et al., 1985]. However, reports of somatostatin reduction or loss in the hippocampus are less consistent; some studies report no change in ATD compared to normals [Ferrier et al., 1983] and others a dramatic reduction in the ATD condition [Davies et al., 1980; Chan-Palay, 1987].

Another peptide, neuropeptide Y (NPY), has also been studied in relation to possible changes in ATD. NPY concentrations are unchanged or even increased in the substantia innominata [Allen et al., 1984]. Studies in the cerebral cortex of ATD patients indicated that the greatest severity of NPY loss parallels histopathological abnormalities shown by plaques and tangles. Of the areas studied, the temporal cortex was most severely involved [Chan-Palay et al., 1985a, b]. NPY axons contribute to neuritic plaque formations [Dawbarn et al., 1984; Chan-Palay et al., 1985b]. The hippocampal regions from the brains of ATD patients have been shown [Chan-Palay et al., 1986a, b] to have afflictions in NPY-immunoreactive (NPY-i) networks in the hilus, CA_1, the parasubiculum and entorhinal cortex. Parallel semiquantitative estimates made of the numbers of neuritic plaques and neurofibrillary tangles in the other hippocampi of the brains in every ATD case showed that the areas of heaviest pathological change by these indices are CA_1 and the entorhinal cortex, the subicular complex CA_3 and the area dentata being less affected. NPY-i axons were shown to participate in neuritic plaques. Thus, areas with the most severe loss of NPY-i neurons and axons, CA_1 and the entorhinal cortex, are synonymous with those areas most severely affected by the other indices of ATD, thus confirming that NPY-i networks are involved in the ATD disease process. However, other NPY-i neurons survive, in some subfields better than others. The cumulative evidence suggests a population of hippocampal peptide neurons that are remarkably resistant in terminal neurological disease. The present study aims to clarify the changes encountered in somatostatin neurons in severe ATD and to compare them with changes of NPY cells and fibers.

Studies of the possible coexistence of neuropeptides in the forebrain and cortex of animals and humans show that NPY and somatostatin (SOM) may be colocalized [Hendry et al., 1984; Köhler et al., 1986b; Chan-Palay et al., 1986b; Chan-Palay, 1987]. We include in this paper studies using double label immunocytochemistry to demonstrate that SOM-i and NPY-i coexist in the hippocampal neurons, and the extent and regional differences inherent in this peptide-peptide coexistence.

The hippocampal region retains its reputation for the presence of an extensive and neurochemically varied neuronal population and extrinsic and intrinsic projections. There is the well-documented catecholamine, serotonin, and cholinergic input from brain stem [Lewis et al., 1967; Fuxe, 1965; Köhler et al., 1981; Lidov, 1980; Swanson and Hartman, 1975] and septal nuclei [Alonso and Köhler, 1984; Amaral and Kurz, 1985; Armstrong et al., 1983; Oderfeld-Nowak et al., 1974; Shute and Lewis, 1967; Mellgren and Srebro, 1973]. Neuropeptides are also well represented with (a) cholecystokinin octapeptide (CCK-8) [Greenwood et al., 1982; Köhler and Chan-Palay, 1982a; Stengaard-Pederson et al., 1983; Loren et al., 1979; Roberts et al., 1984], (b) corticotropin-releasing factor (CRF) [Swanson et al., 1983], (c) SOM [Finley et al., 1981; Köhler and Chan-Palay, 1982b, 1983; Morrison et al., 1985; Roberts et al., 1984; Bakst et al., 1985; Vincent et al., 1985; Chan-Palay, 1987], (d) vasoactive intestinal polypeptide (VIP) [Köhler, 1983; Köhler and Chan-Palay, 1983; Loren et al., 1979; Roberts et al., 1984], (e) substance P [Davies and Köhler, 1985; Gall and Selawski, 1984], (f) dynorphin-A_{1-17} [McGinty et al., 1983; Charkin et al., 1985], NPY in extensive locations in the hippocampus of rats and monkeys [Köhler et al., 1986c] and for man [Chan-Palay et al., 1986a, b].

Antibodies against SOM have made it possible to localize neurons containing this peptide using immunohistochemical techniques. Numerous studies detail the presence of SOM neurons and their pathways in the brain [for review, see Nieuwenhuys, 1985]. The present study focuses on the distribution of SOM in the neurons of the human hippocampus and temporal cortex by using immunocytochemical methods, several anti-SOM antibodies, and the immunogold silver method [Chan-Palay, 1987].

Material and Methods

Case Selection for ATD

The object of this part of the study was to compare the distribution of SOM-i and NPY-i neurons in ATD and age-matched controls in the hippocampus and the temporal cortex. The cases were selected according to clinical and neuropathological criteria established at the outset, prior to collection of the material for this study. The study includes 3 cases of severe ATD and 3 control cases. The control subjects had no clinical history of neurological disease. ATD cases and control subjects were matched for age. The ATD subjects had cases with documented clinical histories of progressive dementia and their therapeutic management included symptomatic treatment and prolonged hospitalization. Neither ATD nor control groups included patients with history of stroke, cerebral tumors or brain-related infections. Cases with evidence of vascular disease at gross neuropathological examination were eliminated. Cases with histories of hypertension were included but not those with history of cerebral infarcts. All 6 cases, ATD and controls, underwent gross neuropathological examination and microscopic examination of the brain. Gross examination of the brain included assessment of cortical atrophy and ventricular dilation. One hemisphere was then removed for neuropathological examination, the other for immunocytochemistry. The question of laterality was avoided for the time being by taking alternating hemispheres in chronological order – first case on the left, next on the right, etc. Microscopic examination for neuropathological assessment included routine hematoxylin and eosin stain, Nissl and Bodian silver stains performed on paraffin-embedded tissue, 10 μm thick. In every case the hippocampus, frontal, parietal and temporal cortex were studied. For this study, quantitative estimates were made of the neuritic plaques and neurofibrillary tangles from the Bodian-stained material. One section of the hippocampus taken at random in the rostrocaudal extent was examined in its seven regions: the hilus, CA_3, CA_1, presubiculum, subiculum, parasubiculum and entorhinal cortex. Neuritic plaques and neurofibrillary tangles were counted in three optical fields at 100 × magnification. The mean was taken for these three fields and recorded for the area. These results were compiled and compared between normal and ATD cases region by region. The 6 cases included in this SOM-i neuron distribution study have been previously reported on in our earlier studies on NPY in various cortical regions [Chan-Palay et al., 1985a, b] and in the hippocampus [Chan-Palay et al., 1986a, b]. These ATD cases were selected for the proven severity of the ATD condition by clinical history and quantitation of numerous plaques and tangles in hippocampus and temporal cortex. Postmortem delays averaged 17 h in the ATD group and 19 h in the controls.

Age Controls

In addition to these, 3 separate cases were selected from younger postmortem brains of patients with no clinical manifestation of neurological disease. Their ages were 43, 43 and 45 years, all were male. The hippocampal regions and temporal cortex from these brains were examined in order to determine whether changes possibly attributable to age could be observed when compared to the older group of control brains age matched to the ATD cases.

Antisera

Four polyclonal SOM-28 antisera (Peninsula Lab., Immunonuclear Corp., and Prof. J. Polak) were used. Liquid-phase absorption controls were conducted by incubation of tissue sections with antisera preadsorbed with SOM-28 (10–100 μg/ml) in diluted serum left for 6 h or overnight for preadsorption. These antigens inactivate the antisera, but glucagon, insulin and vasointestinal polypeptide do not. Adjacent sections were also processed for immunocytochemistry with polyclonal anti-NPY, as previously described [Chan-Palay et al., 1985a, b, 1986a, b], for comparisons between the distributions of SOM-28 and NPY. For the purposes of this study, SOM-28 immunoreactivity will be indicated as SOM-i in the text.

Anti-NPY was raised in rabbits against unconjugated porcine NPY (Cambridge UK Lot *1086). The antibody showed negligible cross-reactivity with neuropeptide PYY and avian pancreatic polypeptide, and no cross-reactivity with other peptides such as SOM and substance P. Incubation of tissue sections with antiserum preadsorbed with 10 µmol of synthetic NPY completely blocks staining of neuronal processes in the hippocampus.

Reactions

Immunocytochemistry was performed on 100-µm thick vibratome slices incubated free floating as previously described [Chan-Palay et al., 1985a, 1986a, b]. Primary SOM antiserum was used at a dilution of 1:1,000 in 0.5% Triton X-100 and 1% normal goat serum (NGS). Subsequent processing was in accordance with the peroxidase-antiperoxidase method [Sternberger, 1979]. Alternate serial sections were used for immunostaining with SOM antibodies and for staining with cresyl violet for the study of cortical lamination. Other sections in a parallel series were visualized using the immunogold/silver staining method [Chan-Palay, 1987].

Double Label with Thioflavin for Neuritic Plaques

Selected sections of hippocampus from ATD cases incubated with SOM antisera were counterstained with 1% thioflavin (Sigma) for 30 min, differentiated in 70% alcohol, dehydrated in ascending alcohol series prior to permanent coverslipping with Eukitt (Bern, Switzerland). These preparations were examined with transmitted light as well as with fluorescence microscopy in order to determine the coexistence of SOM-i elements in neuritic plaques identified by the amyloid stain with thioflavin [Chan-Palay et al., 1985b, 1986b].

Double Labelling: SOM-i:NPY-i Coexistence

Cryostat sections 10 µm thick or vibratome sections not more than 20 µm thick were prepared as pairs of sections. Strict attention was paid to keep track of the adjacent surfaces between the two sections. These exposures of the two adjacent surfaces were achieved by inverting one of the sections. Thus, the exposed surfaces represent the two facing surfaces of one section procedure. For studying coexistence in the light microscope, cells lying in the plane of section could be cut into two parts, one part remaining within the exposed surface of the first section face, and the other part in the second exposed face. Other cells could remain unsectioned in one or the other face, of course. Each section was then treated separately but simultaneously in individual wells with either a SOM or NPY antiserum in order to examine the coexistence of these two immunoreactivities in neurons. The following procedures were used. Pairs of sections were floated onto 1% hydrogen peroxide in phosphate-buffered saline for 2 h in order to reduce or eliminate endogenous peroxidase reactions, reacted sequentially with the primary antibody (1:1,000 NPY or 1:1,500 SOM) overnight, goat anti-rabbit antibodies, peroxidase and diaminobenzidine (DAB) as previously described [Chan-Palay et al., 1985a, b, 1986a, b; Chan-Palay and Yasargil, 1986]. Alternatively the immunogold/silver reaction described in the section above was also used on paired sections with one antibody on each of the pairs. Or, one of a pair of sections was treated with the immunogold/silver method and the other with the peroxidase-antiperoxidase method. In another group of double label experiments, pairs of sections were simultaneously incubated with both antibodies on the same section in a mixture of the polyclonal NPY (1:1,000) and the monoclonal SOM (1:1,500).

Subsequently, fluorescein (FITC) conjugated with goat anti-rabbit immunoglobulins and rhodamine-conjugated anti-mouse immunoglobulins were applied. The preparations were mounted in a glycerin-buffer mixture (1:3) for examination in a bright-field and fluorescence microscope [see Chan-Palay, 1987].

Mapping

In order to localize the neuronal elements accurately in the different hippocampal regions, maps were made of all the specimens, charting the lamination of the regions in relation to stained neurons. Identified SOM-i neurons were counted with attention to their specific location in the hippocampal layers, regions and the major fiber bundles. Because of the difficulty of maintaining consistent section thickness, these counts were considered only semiquantitative. Black and white and color photomicrographs were made on a Zeiss photomicroscope fitted with Nomarski optics. The nomenclature for the hippocampal region follows that used in Schaltenbrand and Bailey [1959], Braak [1974] and Kemper [1984]. For the double staining SOM-i:NPY-i coexistence studies, all neurons immunoreactive to either NPY or SOM from pairs of adjacent peroxidase-stained sections were mapped as previously described Chan-Palay [1987].

Results

Distribution of SOM-i Neurons in the Hippocampus in ATD (fig. 1b, 2, 3a, b)

Each case when carefully examined revealed some differences in the degree of involvement of SOM-i neurons in the ATD process. However, by the criteria of the numbers of neuritic plaques and filamentous tangles, all cases were classified as severe. However, 1 of the 3 cases had hippocampal gyri that were atrophied to almost a third of the normal volume, and with all of the SOM-i neurons no longer persisting in the severely atrophied parts.

In general, ATD material immunostained with SOM antisera is readily distinguished from the control material on the basis of the particular morphology of the individual SOM neurons and networks. SOM-i neurons of the ATD type have a dense, intense immunoreactive character, the deposit of silver enhancement grains is smooth and fills the neuron. Dendrites are foreshortened and axonal fibers are wispy and fragile in appearance, if present at all. The varicosities are larger than in the control material and the intervaricose segments are difficult to follow except for the occasional very thick ones and for short distances. All observed SOM-i neurons were of the nonpyramidal type, multipolar or bitufted in shape and of various sizes. The locations of SOM-i neurons affected by ATD have been charted onto the schema in figure 1b and the black dots indicate

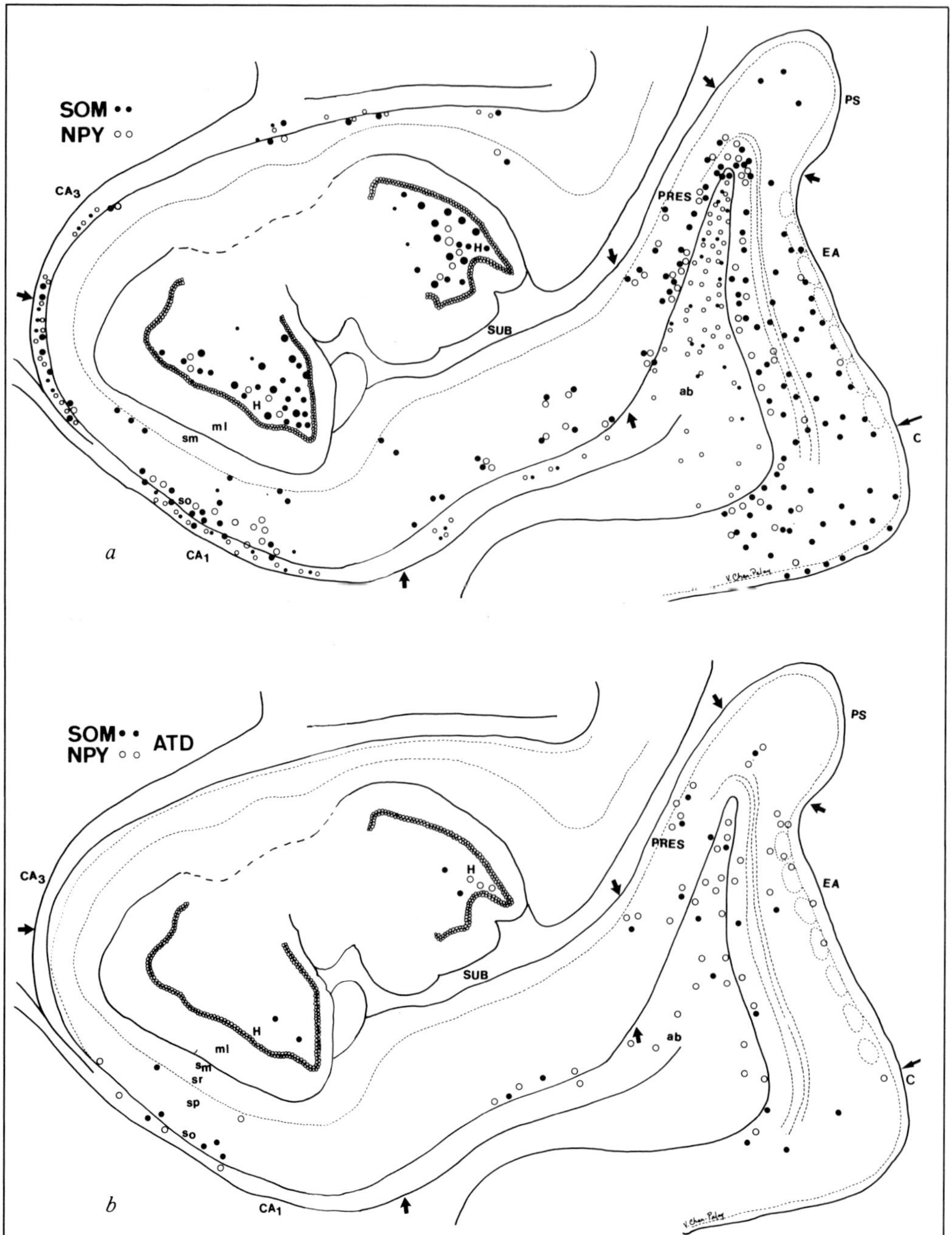
SOM • •
NPY ○ ○
CA3
H
H
ml
sm
so
CA1
SUB
PRES
PS
EA
ab
C
a
SOM • • ATD
NPY ○ ○
CA3
H
H
ml
sm
sr
sp
so
CA1
SUB
PRES
PS
EA
ab
C
b

1

surviving SOM-i ATD neurons with the various subfields. The schema also indicates by open circles the relative number of NPY-i neurons remaining in the same subfields (fig. 1b). The data concerning the NPY-i neurons in ATD have been previously described in detail [Chan-Palay et al., 1986b] and are included here for comparative purposes. Fundamentally, the

---

Fig. 1. a Schematic representation of the rostral hippocampus with sites of SOM-i neurons (black dots in small, medium and large sizes) and NPY-i neurons (open circles) superimposed. Both peptides have comparable distributions and exhibit a significant proportion of SOM-i/NPY-i coexistence (see text). Control brains. Immunogold/silver method. *b* Schematic representation of the rostral hippocampus with sites of SOM-i neurons (black dots) and NPY-i neurons (open circles) superimposed in the ATD brains. Notice the severe loss in SOM-i neurons and the comparably smaller loss in NPY-i neurons (see text). AD = Area dentata; C = perirhinal cortex; CA_1, CA_3 = subfields of Ammon's horn; EA = entorhinal area; H = hilus of area dentata; PRES = presubiculum; PS = parasubiculum; SUB = subiculum; ab = angular bundle; f = fimbria; ml = molecular layer of the area dentata; gl = granular layer of the area dentata; sm = stratum moleculare; so = stratum oriens; sp = stratum pyramidale; sr = stratum radiatum.

Fig. 2. Diagrams to illustrate the differences between appearances of dendrites, cell somata *(a–c)*, and axons *(d–f)* of NPY-i neurons in material from neurosurgical preparations, postmortem controls from aged brains, and ATD brains (for details see text).

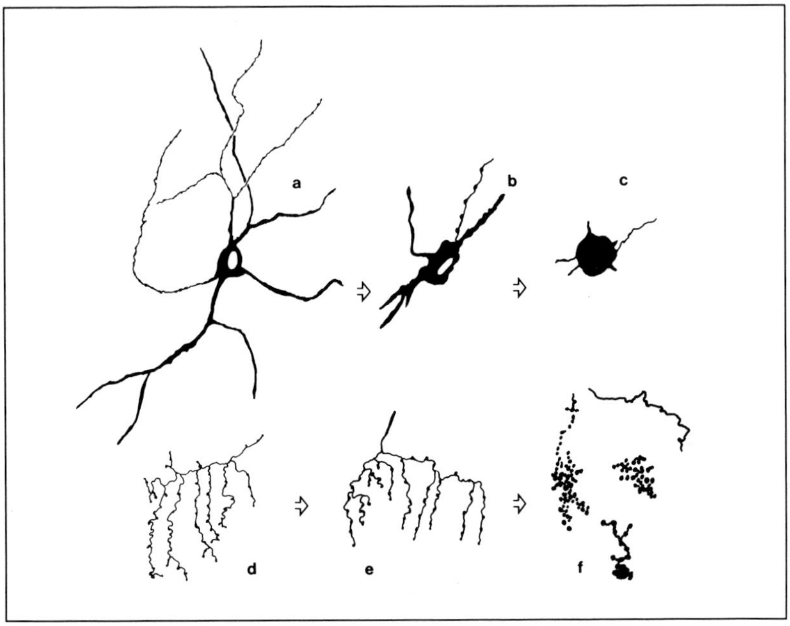

2

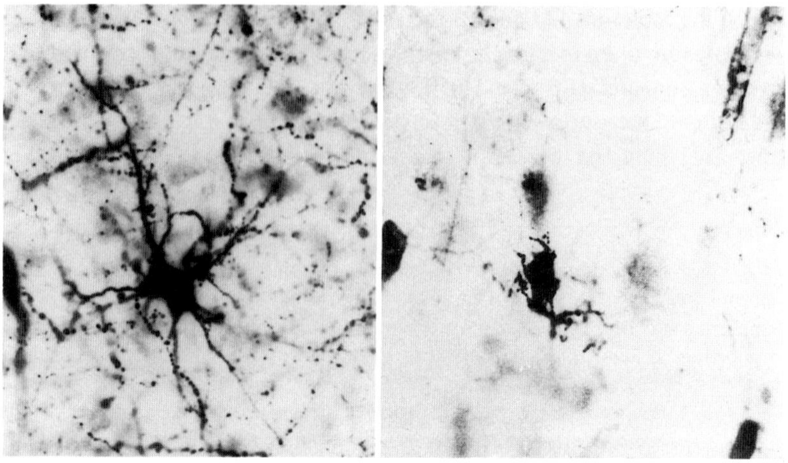

Fig. 3. Small multipolar NPY-i neuron in the entorhinal cortex in a control brain *(a)* compared to a similar NPY-i neuron in this area in the Alzheimer brain *(b).*

whole Ammon's horn, but particularly the area dentata, and the subfield CA_1 are most severely affected by ATD. The few SOM-i neurons which remain are vestiges, with few processes and misshapen perikarya. The rest are gone altogether, a severe regional atrophy has occurred.

In the subicular region ATD has less effect on SOM-i. The entorhinal cortex and perirhinal cortex, however, are sites of serious SOM-i neuron loss. Compared to the controls, the numbers of surviving SOM-i neurons are considerably reduced. There are stratification differences in this reduction in the cortical regions. The upper layers (I, II) are basically devoid of the few SOM-i neurons that might normally be found there. The middle layers have a few neurons, sometimes found in pairs or small groups of middle sized SOM-i cells. In the deepest layers one finds the best survival of SOM-i neurons. This coincides with the fact that the deeper layers are also where the largest numbers of SOM-i neurons occur. The larger SOM-i cells survive well and middle sized cells and a few of the small fusiform bipolar neurons in the white matter also remains.

Comparatively, in the same subfields in these cases of ATD, NPY-i neurons survive in greater numbers than the SOM-i neurons. This is particularly significant as there are more SOM-i neurons as detected by the NPY and SOM antibodies that we have used consistently in these studies.

Nevertheless, one may have to consider that the proportional relationship between numbers of neurons revealed by NPY and by SOM immunoreactivity may reflect the specificities of these two antibodies for their respective antigens under the technical constraints imposed by fixation as much as the biological activities and states of the neurons.

Comparisons between the Distribution of SOM-i and NPY-i Neurons in the Hippocampus and Temporal Cortex (fig. 1a, b)

SOM-i neurons are most abundant in the hilus of the area dentata, in strata oriens and pyramidale in the CA_1 subfields, and in the deeper layers of the entorhinal cortex (II through VI and white matter) and the perirhinal cortex. Small fusiform neurons occur in the white matter, fimbria, alveus and angular bundle, and white matter in the hippocampus, retrohippocampus and temporal cortex. NPY-i neurons are most abundant in the area dentata and the strata oriens and pyramidale of CA_1. The subicular complex, entorhinal cortex and the temporal cortex have numerous NPY-i neurons. Most prominent are the small fusiform neurons of the white matter in fimbria, alveus and the angular bundle [Chan-Palay et al., 1986a, b]. The distributions of these two types of peptide immunoreactivities are remarkably similar, and the superimposition of the two maps compiled from the rostral hippocampus, one showing NPY-i neurons and the other showing SOM-i neurons, supports this impression (fig. 1a). The major similarities of the two groups are those of like cell morphology and cell size and parallel locations of neuron distribution. The fact that neither cell population includes pyramidal neurons in our material is also relevant.

Several minor differences exist as well between these two neuronal immunoreactivities. The SOM-i neuron population by and large in every region of hippocampus and cortex is more numerous than the NPY-i population. This is especially evident in the hilar regions and in the entorhinal and temporal cortex. However, the reverse is true of the small, fusiform neurons in the white matter. Although common to both, these cells have a distinctly higher incidence in NPY-i material.

The parallel distribution of SOM-i and NPY-i cells raises the question of whether these neurons are separated neuron populations, or whether SOM-i and NPY-i coexist in some neurons. If SOM-i/NPY-i coexistence occurs, then to what extent and what are the differences in their regional distributions?

Table I summarizes the SOM and NPY innervations in seven hippocampal regions and the temporal cortex in the control and ATD cases

Table I. SOM and NPY innervations in seven hippocampal regions and temporal cortex control cases and ATD cases[1,2,3]

		Hilus	CA_3	CA_1	Subiculum	Pre-subiculum	Para-subiculum	Entorhinal cortex	Temporal cortex
Alzheimer	SOM	0	0	*	*	*	*	*	*
Control	SOM	***	*	**	**	**	**	***	***
Alzheimer	NPY	0	0	*	*	*	*	*	*
Control	NPY	+	+	+++	++	++	+++	+++	++

[1] Key: 0 = rare SOM neurons; * = SOM neurons; ** = moderate SOM neurons; *** = numerous SOM neurons.
[2] Key: 0 = rare NPY neurons; + = NPY neurons; ++ = moderate NPY neurons; +++ = numerous NPY neurons.
[3] The areas with dramatic reduction of peptide-immunoreactive axons, terminals, and neurons in the diseased hippocampi are the hilar region, CA_1, parasubiculum, and entorhinal cortex and temporal cortex. Less obvious reductions occurred in CA_3, the subiculum, and presubiculum. These findings parallel those in which the same areas have high neuritic plaque or neurofibrillary tangle counts.

Table II. Counts of neuritic plaques/neurofibrillary tangles in 7 hippocampal regions and temporal cortex ATD and age-matched controls[1,2]

	Age	Sex	Hilus	CA_3	CA_1	Subiculum	Pre-subiculum	Para-subiculum	Entorhinal cortex	Temporal cortex
Alzheimer cases										
5	82	F	*/+	0/0	***/+++	**/++	*/+	*/+	**/++	**/+
6	84	M	*/+	*/+	**/+++	*/++	*/+	**/+	***/++	***/+
7	82	F	***/++	*/+	**/++	*/+	0/+	*/++	*/++	**/+
Control cases										
2	72	M	0/0	0/0	0/0	0/0	0/0	0/0	0/0	*/0
3	72	M	0/0	0/0	0/0	0/0	0/0	0/0	0/0	0/0
4	83	M	0/0	0/0	0/0	0/0	0/0	0/0	0/0	0/0

[1] Neuritic plaques: 0 = minimum plaques (0–1); * = few plaques (2–10); ** = moderate plaques (11–20); *** = numerous plaques (> 20). Neurofibrillary tangles (NFT): 0 = minimum NFT (0–1); + = few NFT (2–10); ++ = moderate NFT (11–20); +++ = numerous NFT (> 20).
[2] The regions are the hilus, CA_3, CA_1, subiculum, presubiculum, parasubiculum, and entorhinal cortex. The counts are highest in CA_1 and entorhinal regions, and less dramatic in the subicular complex and the hilus.

selected for this study. The subfields with the most severe loss in SOM-i and NPY-i neurons are the hilus of the area dentata, CA_1 and the entorhinal, perirhinal and temporal cortices although SOM loss is more severe than the NPY. Table II summarizes the details of the counts for neuritic plaques and neurofibrillary tangles in the seven hippocampal and temporal cortex in the same 3 ATD cases and 3 controls. The area with the highest densities of plaques and tangles are the hilus of the area dentata, CA_1 and the entorhinal and temporal cortices. The numbers of neuritic plaques and neurofibrillary tangles reported in these cases are in accordance with the guidelines recently established for the diagnosis of ATD [Katchaturian, 1985], and CA_1 and the entorhinal cortex are known to be sites of predilection of ATD morphological changes [Tomlinson and Corsellis, 1984; Ball, 1978; Hooper and Vogel, 1976; Wilcock and Esiri, 1982; Hyman et al., 1984].

Thus, the present study clearly indicates that in severe ATD, the predilection sites for highest incidence of neuritic plaques and neurofibrillary tangles are the sites of the severest loss of SOM-i neurons and networks. These regions are also the sites of the most severe loss and alterations in NPY-i neurons and networks [Chan-Palay et al., 1985a, b; 1986a, b], implicating both peptide neuron systems in the disease process.

Coexistence between SOM-i and NPY-i in Single Neurons (fig. 4a–c)

The surgical biopsy material consistently provided the most reliable tissues for the double labelling experiments to demonstrate coexistence of SOM-i and NPY-i in neurons, compared to human postmortem material, regardless of postmortem delay or conditions of perfusion and fixation. The major factor was the clearly superior preservation and fixation consistently achievable. Vast numbers of axons are present and dendrites are fully stained. This experience corroborates similar observations from the demonstration of NPY-i in human cortical tissue in surgical and postmortem tissues [Chan-Palay and Yasargil, 1986]. The parallel distributions of SOM-i and NPY- neurons (fig. 1a) underscore the impression that these two peptides occur in similar neuronal populations in comparable subfields of the hippocampus and cortex. The results of the double labelling experiments demonstrate that about 30% of the SOM-i neurons in the hilus of the area dentata are also NPY-i. However, since numerically more SOM-i neurons are visualized by our methods than NPY-i neurons, the coexistence population constitutes approximately 50% of all NPY-i neurons. NPY-i coexistence was found in several SOM-i neurons in CA_1, the

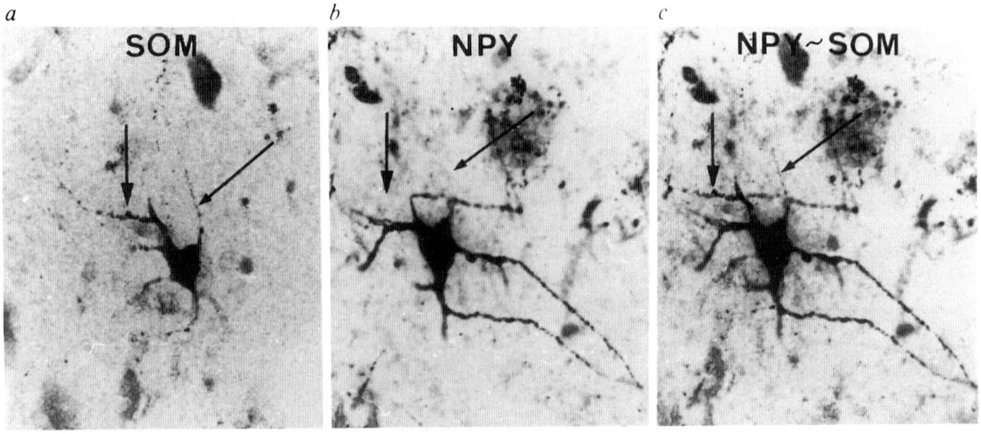

Fig. 4. Coexistence of NPY-i and SOM-i in a single large multipolar neuron of the stratum oriens at the CA_1/subicular border *(a–c).* SOM-i by the immunogold/silver method and NPY-i is demonstrated by the peroxidase-antiperoxidase preparation method *(b)* on the adjacent section. The labelled neuron shown in these two figures comes from the adjacent opposing faces of two sections. The two faces are superimposed *(c).* × 500.

dentate hilus and particularly in the subicular complex. The highest coexistence coefficients were found in the deep layers of the entorhinal and perirhinal cortices where more than 50% of the SOM-i neurons were NPY-i as well. Large neurons of the retrohippocampal areas and small neurons in the white matter were particularly noted as frequent examples of NPY-i and SOM-i coexistence.

In the ATD cases, both NPY-i and SOM-i in neurons and axons were seriously affected with a numerically higher survival of NPY-i neurons and axons than SOM-i neurons and axons in the cortical areas, and equally severe destruction of both peptide neurons in the area dentata and CA_1. Neurons with coexistence of NPY-i and SOM-i appear not to be better protected against ATD than the neurons with SOM-i alone, and remain or are destroyed in the same poportion as those with SOM-i alone. Because of the techniques that we use for the demonstrations of coexistence, we anticipate that our estimates are generally lower than might otherwise be found. Thus, the numerical estimates given here can be considered minimal.

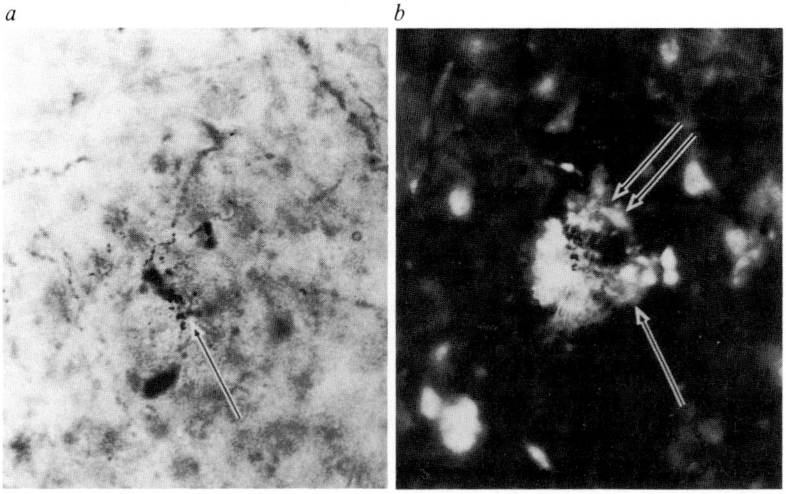

Fig. 5. NPY-i axonal clusters are colocalized with neuritic plaques formations by NPY immunocytochemistry and thioflavin counterstain. *a* A small NPY-i axonal cluster (arrow) identified by immunocytochemistry and bright-field microscopy. *b* The same NPY-i cluster (arrow) shown in center of a large neuritic plaque (double arrows) identified by thioflavin in a double exposure, fluorescence and bright-field micrograph. ×400.

In order to demonstrate the relationship of SOM-i neurons to neuritic plaques, a double label experiment was initiated with thioflavin after the silver enhancement procedure (see also fig. 5a, b). The amyloid within the neuritic plaques is well visualized by fluorescence microscopy. In these preparations, when SOM-i elements are present and are well enhanced by the immunogold/silver method, both labeled neurons and neuritic plaques are visualized in the same section with fluorescence and transmitted light. In the hippocampal gyri where the cortex is completely 'burned out', even neuritic plaques cannot be found, and no SOM-i neurons are seen. In the better preserved sections of the entorhinal and perirhinal cortices where the ATD process is less severe, neuritic plaques are readily found. Although SOM-i axons within plaques are readily recognized, as previously reported [Armstrong et al., 1985; Nakamura and Vincent, 1986], we add the observation that SOM-i neurons/perikarya do not occur within the neuritic plaques itself, despite the fact that well-preserved labeled neurons appear on the periphery of neuritic plaques. I have no information on the coexistence of SOM-i neurons and neurofibrillary tangles.

Discussion

The SOM system in the human hippocampus and retrohippocampal regions is strikingly similar to that of NPY. While SOM neurons are found throughout the regions studied, the highest densities of SOM-i neurons occurred in the hilus of the area dentata, the CA_1 region, and in the deep layers of the retrohippocampal cortices – entorhinal and perirhinal. Axonal networks are comparably more complicated and dense in these same areas. This remarkable comparability of the exaggerated densities of SOM-i neuronal population and of SOM-i axonal populations in these particular regions suggests the likelihood that the SOM-i neurons give rise to much of the dense axonal networks in their immediate subfield, indicating that a major part of the SOM-i neurons population are interneuronal and that much of the SOM-i fibers in these regions are part of these SOM-i interneuronal circuits. This does not rule out the possibility that hippocampal sources of SOM-i fibers exist.

A major application of this knowledge of the SOM-i circuits in the human hippocampal and retrohippocampal regions is provided by the ensuing study of the 3 cases of ATD which were selected for (a) the severity of their clinical conditions, (b) the high incidence by counts of neuritic plaques and neurofibrillary tangles, and (c) the demonstrated severity in loss and alterations in NPY-i neuronal circuits demonstrates that there is a parallel severe loss of SOM-i neurons and axonal networks and serious alterations in the SOM-i innervation of these regions. These results are consistent with other observations indicating that there is severe loss of measurable SOM in hippocampus and temporal cortex in severe ATD [Davies et al., 1980; Rossor et al., 1982; Ferrier et al., 1983; Roberts et al., 1985] and disagrees with others that claim that the only visible change to SOM-i neurons is some axon swelling [Nakamura and Vincent, 1986]. The selective destruction of the SOM-i system parallels the severity of ATD as recorded by histopathological alterations of plaques and tangles in the various hippocampal subfields, as well as the severity of the disease as expressed in the loss and alterations observed in NPY-i circuits in the same hippocampal regions [Chan-Palay et al., 1986a, b]. Thus, both peptide systems are involved in the disease.

The present study has also provided fundamental information on the phenomenon of peptide-peptide coexistence in human neurons. The extent of this distribution and the regional differences present in the various hippocampal subfields is also discussed. The importance of this observation is the fact that populations of SOM-i/NPY-i neurons exist in considerable numbers. Neurons with SOM-i and without NPY-i also exist as do their opposite counterparts. There is equally severe destruction of both peptide neuron groups in the area dentata and CA_1; although a higher incidence of NPY-i neurons and networks survive in the cortices compared to SOM-i neurons.

One would expect that this loss of a peptidergic neuron network would be important to the functioning of the hippocampal/limbic circuits perhaps as reflected in severely decompensated state of these patients typically. Nevertheless, a more significant observation may be that a considerable NPY-i intrinsic network of cells and fibers continues to exist into the terminal state at exitus. This unusual survivability points to the fact that these peptide neurons have participated in the maintenance of the neuronal networks required for whatever minimal function remains in the target area of the terminally diseased brain. As such they hold important links for our understanding of synaptic plasticity in neural reorganization and repair. This observation is true for NPY-i neuron networks not only in the hippocampus but in the cerebral cortex in ATD [Chan-Palay et al., 1985a, b], in the striatum in Huntington's disease [Allen et al., 1984; Dawbarn et al., 1984], and in the region directly bordering on brain infarcts [Chan-Palay and Yasargil, 1986]. The cumulative evidence suggests an NPY population throughout the hippocampus remarkably resistant to Alzheimer changes and a particularly susceptible SOM one.

Conclusion and Summary

This paper provides detailed information on the distribution of NPY-i and SOM-i neurons and fibers in the hippocampal region of neuropathologically confirmed cases of ATD at postmortem compared to normal aged-matched controls. These neuronal networks are detected by polyclonal antibodies and controls were obtained by using liquid-phase absorption immunocytochemistry. The description covers the subfields area dentata, CA_3 and CA_1, the subicular complex and entorhinal area. The hippocampal regions with the most severe afflictions in NPY-i neuron networks are the hilus, CA_1, the parasubiculum and the entorhinal cortex. Less obvious reductions occurred in CA_3, subiculum and the presubiculum. Parallel semiquantitative estimates were made of the numbers of neuritic plaques and neurofibrillary tangles in the hippocampus of the brains in every ATD case. The areas of heaviest pathological changes by these indices are CA_1 and the entorhinal cortex. The subicular complex CA_3 and the area dentata are less affected. These findings show that the areas with the most severe loss of NPY-i neurons and axons, CA_1 and the entorhinal cortex, are synonymous in location with those areas most severely affected by the other indices of ATD. Thus, NPY-i networks are involved in the ATD disease process. However, other NPY-i networks survive, in some subfields better than others. The most prominent accumulations of SOM-i neurons and axons occurred in the hilus of the area dentata, in CA_1, and in the entorhinal, perirhinal cortices. Axonal plexuses occurred throughout the hippocampal subfields but were particularly dense in those regions rich in SOM neurons. The distribution of SOM-i neurons and fibers parallels the distribution of NPY neurons and fibers in the hippocampus and cerebral cortex to a remarkable extent. Double labelling experiments with antisera against NPY and SOM indicate a considerable level of coexistence of the two peptides in single neurons, particularly in large multipolar cortical neurons and also in the small bipolar white matter neurons. Regional variations exist in the amounts of coexistence found in the hippocampal subfields. SOM-NPY coexistence is particularly high in the hilus of the area dentata, the subicular complex, and the deep layers of the entorhinal and perirhinal cortices. In the hippocampi and temporal cortices of ATD, SOM-i neurons and axons show a severe loss, which is worst in those regions with the highest indices of neurofibrillary tangles and neuritic plaques – the hilus of the area dentata, CA_1, and the entorhinal and perirhinal cortices. Neuritic plaques identified on double label experiments with thioflavin include SOM axons but not neurons.

References

Allen, J.M.; Fernier, I.N.; Roberts, G.W.; Cross, A.J.; Adrian, T.E.; Crow, T.J.; Bloom, S.R.: Elevation of neuropeptide Y in substantia innominata in Alzheimer's type dementia. J. neurol. Sci. *64:* 325–331 (1984).

Alonso, A.; Köhler, C.: A study of the reciprocal connections between the septum and the entorhinal area using anterograde and retrograde axonal transport methods in the rat brain. J. comp. Neurol. *225:* 327–343 (1984).

Amaral, D.G.; Kurz, J.: An analysis of the origins of the cholinergic and noncholinergic septal projections to the hippocampal formation of the rat. J. comp. Neurol. *240:* 37–60 (1985).

Armstrong, D.M.; LeRoy, S.; Shields, D.; Terry, R.D.: Somatostatin-like immunoreactivities within neuritic plaques. Brain. Res. *338:* 71–79 (1985).

Armstrong, D.M.; Saper, C.B.; Levey, A.I.; Wainer, B.M.; Terry, R.D.: Distribution of cholinergic neurons in rat brain demonstrated by the immunocytochemical localization of choline acetyltransferase. J. comp. Neurol. *216:* 53–68 (1983).

Bakst, I.; Morrison, J.H.; Amaral, D.G.: The distribution of somatostatin-like immunoreactivity in the monkey hippocampal formation. J. comp. Neurol. *236:* 423–442 (1985).

Ball, M.J.: Topographic distribution of neurofibrillary tangles and granulovacuolar degeneration in hippocampal cortex of aging and demented patients. Acta neuropath. *42:* 73–80 (1978).

Blessed, G.; Tomlinson, B.E.; Roth, M.: The association between quantitative measurements of dementia and senile changes in the cerebral grey matter of elderly subjects. Br. J. Psychiat. *114:* 797–811 (1968).

Bowen, D.M.; Smith, C.B.; White, B.; Davison, A.N.: Neurotransmitter related enzymes and indices of hypoxia in senile dementia and other abiotrophies. Brain *99:* 459–466 (1976).

Braak, H.: On the structure of the human archicortex. Cell Tiss. Res. *152:* 349–383 (1974).

Carlsson, A.; Adolfsson, R.; Aquilonius, S.M.; Gottfries, C.; Oreland, L.; Svennerholm, L.; Winblad, B.: Biogenic amines in human brain in normal aging, senile dementia and chronic alcoholism; in Goldstein, Ergot compounds and brain functions: neuroendocrine and neuropsychiatric aspects, pp. 295–304 (Raven Press, New York 1980).

Chan-Palay, V.: Somatostatin immunoreactive neurons in the human hippocampus and cortex shown by immunogold/silver intensification on vibratome section: coexistence with neuropeptide Y neurons, and effects in Alzheimer's type dementia. J. comp. Neurol. (in press, 1987).

Chan-Palay, V.; Allen, Y.S.; Lang, W.; Haesler, U.; Polak, J.M.: I. Cytology and distribution in normal human cerebral cortex of neurons immunoreactive with antisera against neuropeptide Y. J. comp. Neurol. *238:* 382–390 (1985a).

Chan-Palay, V.; Köhler, C.; Haesler, U.; Lang, W.; Yasargil, G.: Distribution of neurons and axons immunoreactive with antisera against neuropeptide Y in the normal human hippocampus. J. comp. Neurol. *248:* 360–375 (1986a).

Chan-Palay, V.; Lang, W.; Allen, Y.S.; Haesler, U.; Polak, J.M.: II. Cortical neurons immunoreactive with antisera against neuropeptide Y are altered in Alzheimer's-type dementia. J. comp. Neurol. *238:* 382–390 (1985b).

Chan-Palay, V.; Lang, W.; Haesler, U.; Köhler, C.; Yasargil, G.: Distribution of altered hippocampal neurons and axons immunoreactive with antisera against neuropeptide Y in Alzheimer's type dementia. J. comp. Neurol. *248:* 376–394 (1986b).

Chan-Palay, V.; Yasargil, G.: Immunocytochemistry of human brain tissue with a polyclonal antiserum against neuropeptide Y. Anat. Embryol. *174:* 27–33 (1986).

Charkin, C.; Shoemaker, W.J.; McCinty, J.F.; Bayon, A.; Bloom, F.E.: Characterization of the prodynorphin and proenkephalin neuropeptide systems in rat hippocampus. J. Neurosci. *5:* 808–815 (1985).

Coyle, J.T.; Price, D.L.; Delong, M.R.: Alzheimer's disease: a disorder of cortical cholinergic innervation. Science *219:* 1184–1190 (1983).

Cross, A.J.; Crow, T.J.; Johnson, J.A.; Perry, E.K.; Blessed, G.; Tomlinson, B.E.: Monoamine metabolism in senile dementia of Alzheimer's type. J. neurol. Sci. *60:* 383–392 (1983).

Davies, P.; Katzmann, R.; Terry, R.D.: Reduced somatostatin-like immunoreactivity in cases of Alzheimer's disease and Alzheimer senile dementia. Nature, Lond. *288:* 279–280 (1980).

Davies, P.; Maloney, A.J.: Selective loss of central cholinergic neurons in Alzheimer's disease. Lancet *ii:* 1043 (1976).

Davies, P.; Terry, R.D.: Cortical somatostatin-like immunoreactivity in cases of Alzheimer's disease and senile dementia of Alzheimer type. Neurobiol. Aging *2:* 9–14 (1981).

Davies, S.; Köhler, C.: The substance P innervation of the hippocampus in the rat. Anat. Embryol. *173:* 45–52 (1985).

Dawbarn, D.; Hunt, S.P.; Emson, P.C.: Neuropeptide Y: regional distribution, chromatographic characterization and immunohistochemical demonstration in post-mortem human brain. Brain Res. *296:* 168–173 (1984).

Ferrier, I.N.; Cross, J.A.; Johnson, J.A.; Robers, G.W.; Crow, T.J.; Carsellis, J.A.N.; Lee, Y.C.; O'Shangoressey, A.T.E.; McGregor, G.P.; Bacarese-Hamilton, A.J.; Bloom, S.R.: Neuropeptides in Alzheimer-type dementia. J. neurol. Sci. *62:* 152–170 (1983).

Finley, J.C.W.; Maderdrut, J.L.; Roger, L.J.; Petrusz, P.: The immunocytochemical localization of somatostatin-containing neurons in the rat CNS. Neuroscience *6:* 2173–2192 (1981).

Fuxe, K.: Evidence for the existence of monoamine terminals in the central nervous system. Acta physiol. scand. *64:* 39–85 (1965).

Gall, C.; Selawski, L.: Supramammillary efferents to guinea pig hippocampus contain substance P-like immunoreactivity. Neurosci. Lett. *51:* 171–176 (1984).

Gottfries, C.G.; Adolfsson, R.; Aquilonius, S.M.; Carlsson, A.; Ecernas, S.A.; Nordberg, A.; Oveland, L.; Svennerholm, L.; Wiberg, A.; Winblad, B.: Biochemical changes in dementia disorders of Alzheimer type (AD/SDAT). Neurobiol. Aging *4:* 261–271 (1983).

Greenwood, R.S.; Godar, S.; Reaves, T.A.; Hayward, J.N.: Cholecystokinin in hippocampal pathways. J. comp. Neurol. *203:* 335–351 (1982).

Hardy, J.; Adolfsson, R.; Alafusoff, I.; Bucht, G.; Marcosson, J.; Nyberg, P.; Perdahl, E.; Wester, P.; Winblad, P.: Transmitter deficits in Alzheimer's disease. Neurochem. int. *7:* 545–562 (1985).

Hendry, S.H.C.; Jones, E.G.; Emson, P.C.: Morphology, distribution and synaptic relations of somatostatin and neuropeptide Y-immunoreactive neurons in rat and monkey neocortex. J. Neurosci. *4:* 2497–2517 (1984).

Henke, H.; Lang, W.: Cholinergic enzymes in neocortex, hippocampus and basal forebrain of non-neurological and senile dementia of Alzheimer-type patients. Brain Res. *267:* 281–291 (1983).

Hooper, W.M.; Vogel, F.S.: The limbic system in Alzheimer's disease. Am. J. Path. *85:* 1–19 (1976).

Hyman, B.T.; Hoesen, G.W. van; Damasio, A.R.; Barnes, C.L.: Alzheimer's disease: cell-specific pathology isolates the hippocampal formation. Science *225:* 1168–1170 (1984).

Katchaturian, Z.S.: Diagnosis of Alzheimer's disease. Archs Neurol. *42:* 1097–1105 (1985).

Kemper, J.: Neuroanatomical and neuropathological changes in normal aging and dementia; in Albert, Clinical neurology of aging, pp. 9–52 (Oxford University Press, London 1984).
Köhler, C.: A morphological analysis of vasoactive intestinal polypeptide (VIP)-like neurons in the area dentata of the rat brain. J. comp. Neurol. *221:* 247–262 (1983).
Köhler, C.; Chan-Palay, V.: The distribution of cholecystokinin-like immunoreactive neurons and nerve terminals in the retrohippocampal region in the rat and guinea pig. J. comp. Neurol. *210:* 136–146 (1982a).
Köhler, C.; Chan-Palay, V.: Somatostatin-like immunoreactive neurons in the hippocampus: an immunocytochemical study in the rat. Anat. Embryol. *167:* 151–172 (1982b).
Köhler, C.; Chan-Palay, V.: Somatostatin and vasoactive intestinal polypeptide-like immunoreactive cells and terminals in the retrohippocampal region of the rat brain. Anat. Embryol. *167:* 151–172 (1983).
Köhler, C.; Chan-Palay, V.; Steinbusch, H.W.M.: The distribution and orientation of serotonin fibers in the entorhinal and other retrohippocampal areas: an immunohistochemical study with anti-serotonin antibodies in the rat's brain. Anat. Embryol. *161:* 237–264 (1981).
Köhler, C.; Eriksson, L.; Davies, S.; Chan-Palay, V.: NPY and somatostatin coexist in neurons of the hippocampus (in press, 1986b).
Köhler, C.; Smialovska, M.; Ericsson, L.G.; Chan-Palay, V.: Origin of neuropeptide Y in the rat retrohippocampal region. Neurosci. Lett. *65:* 287–292 (1986c).
Lewis, P.R.; Shute, C.C.D.; Silver, A.: Confirmation from choline acetyltransferase of a massive cholinergic innervation to the rat hippocampus. J. Physiol., Lond. *191:* 215–224 (1967).
Lidov, M.G.W.; Grzanna, R.; Molliver, M.E.: The serotonin innervation of the cerebral cortex. An immunohistochemical analysis. Neuroscience *5:* 207–227 (1980).
Loren, I.; Alumets, J.; Hakanson, R.; Sundler, F.: Immunoreactive pancreatic polypeptide (PP) occurs in the central and peripheral nervous system: Preliminary immunocytochemical observations. Cell Tiss. Res. *200:* 179–186 (1979).
McGeer, P.I.; McGeer, E.G.; Suzuki, J.; Dolman, C.E.; Nagai, T.: Aging, Alzheimer's disease and the cholinergic system of the basal forebrain. Neurology *34:* 741–745 (1984).
McGinty, J.F.; Hendriksen, S.J.; Goldstein, A.; Terenius, L.; Bloom, F.E.: Dynorphin is contained within hippocampal mossy fibers: immunochemical alterations after kainic acid administration and colchicine-induced neurotoxicity. Proc. natn. Acad. Sci. USA *80:* 589–593 (1983).
Mellgren, S.T.; Srebro, B.: Changes in acetylcholinesterase and distribution of degenerating fibers in the hippocampal region after septal lesions in the rat. Brain Res. *52:* 19–36 (1973).
Morrison, J.H.; Rogers, J.; Scherr, S.; Benoit, R.; Bloom, F.E.: Somatostatin in neuritic plaques of Alzheimer's patients. Nature, Lond. *314:* 90–92 (1985).
Nakamura, S.; Vincent, S.R.: Somatostatin and neuropeptide Y-immunoreactive neurons in the neocortex in senile dementia of Alzheimer's type. Brain Res. *370:* 11–20 (1986).
Nieuwenhuys, R.: Chemoarchitecture of the brain, pp. 77–81 (Springer, Stuttgart 1985).

Oderfeld-Nowak, B.; Narkiewicz, O.; Bialowas, J.; Dabrowska, J.; Wieraszko, A.; Bradkowska, M.: The influence of septal nuclei lesions on activity of acetylcholinesterase in the hippocampus of the rat. Acta Neurobiol. exp. *34:* 787–789 (1974).

Perry, E.K.; Perry, R.H.; Blessed, G.; Tomlinson, B.E.: Necropsy evidence of central cholinergic deficits in senile dementia. Lancet *i:* 189 (1977).

Perry, E.K.; Tomlinson, B.E.; Blessed, G.; Bergman, K.; Gibson, P.H.; Perry, R.H.: Correlation of cholinergic abnormalities with senile plaques and mental test scores in senile dementia. Br. med. J. *ii:* 1457–1459 (1978).

Roberts, G.W.; Crow, T.W.; Polak, J.M.: Location of neuronal tangles in somatostatin neurons in Alzheimer's disease. Nature, Lond. *314:* 92–94 (1985).

Roberts, G.W.; Woodhams, P.L.; Polak, J.M.; Crow, T.J.: Distribution of neuropeptides in the limbic system of the rat. The hippocampus. Neuroscience *11:* 35–77 (1984).

Rossor, M.N.; Svendsen, C.; Hunt, S.P.; Mountjoy, C.Q.; Roth, M.; Iversen, L.L.: The substantia innominata in Alzheimer's disease: a histochemical and biochemical study of cholinergic marker enzyme. Neurosci. Lett. *28:* 217–222 (1982).

Schaltenbrand, G.; Bailey, P.: Introduction to stereotaxis with an atlas of the human brain, vol. I–III (Thieme, Stuttgart 1959).

Shute, C.C.D.; Lewis, P.R.: The ascending cholinergic reticular systems: neocortical, olfactory, and subcortical projections. Brain *90:* 497–521 (1967).

Stengaard-Pedersen, K.; Fredens, K.; Larsson, L.I.: Comparative localization of enkephalin and cholecystokinin immunoreactivities and heavy metals in the hippocampus. Brain Res. *273:* 81–96 (1983).

Swanson, L.W.; Hartman, B.K.: The central adrenergic system. An immunofluorescence study of the localization of cell bodies and their efferent connections in the rat utilizing dopamine-β-hydroxylase as a marker. J. comp. Neurol. *163:* 487–506 (1975).

Swanson, L.W.; Sawchenko, P.E.; Rivier, J.; Vale, W.W.: Organization of ovine corticotropin-releasing factor immunoreactive cells and fibers in the rat brain: an immunohistochemical study. Neuroendocrinology *36:* 165–186 (1983).

Tomlinson, B.E.; Blessed, G.; Roth, M.: Observations on the brains of demented old people. J. neurol. Sci. *II:* 205–242 (1970).

Tomlinson, B.E.; Corsellis, J.A.N.: Ageing and the dementias: in Hume-Adams, Corsellis, Duchen, Greenfield's neuropathology, pp. 951–1025 (Arnold, London 1984).

Vincent, S.R.; McIntosh, C.H.S.; Buchan, A.M.J.; Brown, J.C.: Central somatostatin revealed with monoclonal antibodies. J. comp. Neurol. *238:* 169–186 (1985).

Whitehouse, P.J.; Price, D.L.; Struble, R.G.; Clark, A.W.; DeLong, M.R.: Alzheimer's disease and senile dementia: loss of neurons in the central forebrain. Science *215:* 1237 (1982).

Wilcock, G.K.; Esiri, M.M.: Plaques, tangles and dementia: a quantitative study. J. neurol. Sci. *56:* 343–356 (1982).

Victoria Chan-Palay, MD, Neurology Clinic, University Hospital,
Frauenklinikstrasse 26, CH–8091 Zürich (Switzerland)

Interdiscipl. Topics Geront., vol. 25, pp. 59–73 (Karger, Basel 1988)

Subcortical Neuronal Projections to the Cerebral Cortex: Their Role in SDAT

Brigitte Berger, Patricia Gaspar

INSERM U.106, Bâtiment de Pédiatrie, Hôpital Salpêtrière, Paris, France

Introduction

There are two ways of considering the physiopathological importance of lesions in dementia of the Alzheimer type (AD): (1) is there a primary focus of the 'degenerative' process? (2) how do lesions contribute to the memory disturbances and the mental deficiency?

While correlations between the degree of dementia and the cortical alterations (frequency of senile plaques or of neurofibrillary tangles) have been known for long (16), only recent progress in neuroanatomical and biochemical methods led to the discovery of the aminergic and cholinergic projections to the cerebral cortex, and their involvement in AD. These lesions, especially those of the cholinergic system appeared as such a prominent feature that some investigators proposed the hypothesis of an 'isodendritic core disease'. In this concept of AD the emphasis was put on the damage to a certain class of phylogenetically older neurons, extending from the brainstem to the basal forebrain and including the locus coeruleus, the raphe nuclei, the substantia nigra and the substantia innominata (23). The hypothesis that this apparently cortical dementia was mainly of subcortical origin, opened up an intensive research in many related clinical, anatomical and experimental behavioral fields. It was hoped that the predominant involvement of one neurotransmitter system would be demonstrated and the AD might thus become accessible to substitutive therapies similarly to Parkinson's disease. For some years, the ascending cholinergic system has been considered as the possible primary focus of AD, in spite of difficulties to correlate the intensity of specific chemical lesions with the severity of mental impairment. As will be discussed in this contribution, there are reasons to doubt that such a single lesion could cause the entire clinical and pathological spectrum of alterations encountered in AD. Rather, the deficits of several neurotransmitters might contribute, to a various extent, to a complex pathophysiological process.

There are many pitfalls in analyzing and interpreting the anatomical and biochemical alterations of the subcortico-cortical pathways, some of which will be discussed in this volume by C.Q. Mountjoy. In this article, we will concentrate on those changes related to the involvement of subcortical systems projecting to the cortex, which can be followed by morphological means.

Ascending Cholinergic System

Anatomy [reviewed in 15]

The basal forebrain cholinergic neurons, also known as the 'basal magnocellular complex', are a group of contiguous nuclei which include the nucleus basalis of Meynert (NBM), the nucleus of the diagonal band of Broca (NDB), and the medial septal nucleus (MSN). Histochemical demonstration of acetylcholinesterase (AChE) and more specifically of choline acetyltransferase (CAT) has now provided evidence for the cholinergic nature of the majority of the large neurons in these nuclei in man. In the macaque, tracing studies have shown that the NBM provides a cholinergic innervation to the entire cortical mantle and to the amygdala with a precise corticotopic organization, while the MSN and the NDB project essentially to the hippocampal formation. In spite of some overlap in these projections, there is no evidence for an extensive collateralization of the individual NBM axons quite contrary to the widespread and extensively collateralized projections from the locus coeruleus for example. Contrary to its wide output, the NBM is under the control of only a few structures: its cortical feedback comes from a few 'limbic-related' cortical areas and from the amygdala and it also receives afferents from various brainstem nuclei.

As the most extensive lesions of the NBM in the macaque led only to 60–70% decrease of neocortical CAT, the cholinergic input to the cortex must also come from different sources, possibly even from cholinergic neurons within the cortex itself. A cholinergic projection from the brainstem (dorsolateral tegmental nucleus in the pons) to the cortex has also been identified. Some regional differentiation of the cholinergic innervation to the cortex has been suggested: higher levels of CAT were found in the hippocampus and cingulate cortex relative to the neocortical fields in man. There are also indications for a laminar selectivity of this innervation, which predominates in layers I, IV–VI.

Evidence for an Involvement of Cholinergic Pre- and Postsynaptic Markers in Cortical Terminal Fields

The cholinergic deficiency in Alzheimer's disease (AD), first demonstrated in 1976 and confirmed in numerous publications, has been diversely estimated [see references in 9]. These estimates were based on the reductions of CAT activity demonstrable in the neocortex, the amygdala and the hippocampus. The topographical pattern of the cholinergic loss in the cortex usually parallels morphological alterations as will be discussed in more detail by C.Q. Mountjoy in this volume.

Loss of AChE probably reflects the loss of cholinergic neurons, although this enzyme is also associated to noncholinergic, noradrenergic and dopaminergic neurons. Histochemical demonstration of axons with AChE activity has shown a severe reduction of their number in the cortex of patients with AD [2, 9]. Furthermore distorted cholinergic axons participating in the formation of the senile plaques (SP) were demonstrated by histochemical and immunohistochemical stains of AChE (fig. 1a) and CAT [2] respectively.

Contrary to the converging data on the loss of the presynaptic cholinergic markers, reports on the state of cholinergic receptors in AD have been discordant [5, 9, 21]. The majority of the authors has reported normal levels of muscarinic binding sites and a seemingly normal distribution pattern on autoradiograms even in severely affected brain tissue. Other workers have reported some alterations of the muscarinic binding sites, but these were moderate changes, restricted to few limbic-related areas. Some authors have indicated that only the high affinity receptor sites M2 were affected, and provided experimental evidence for a presynaptic localization of these receptor sites [21]. These findings were not confirmed by other studies who stated that the proportion of high and low affinity binding sites was undisturbed. On the other hand, the normal or reduced postsynaptic binding sites for AChE have been considered to reflect the loss or alteration of the cortical cholinoceptive neurons, which would fail to express a denervation supersensitivity.

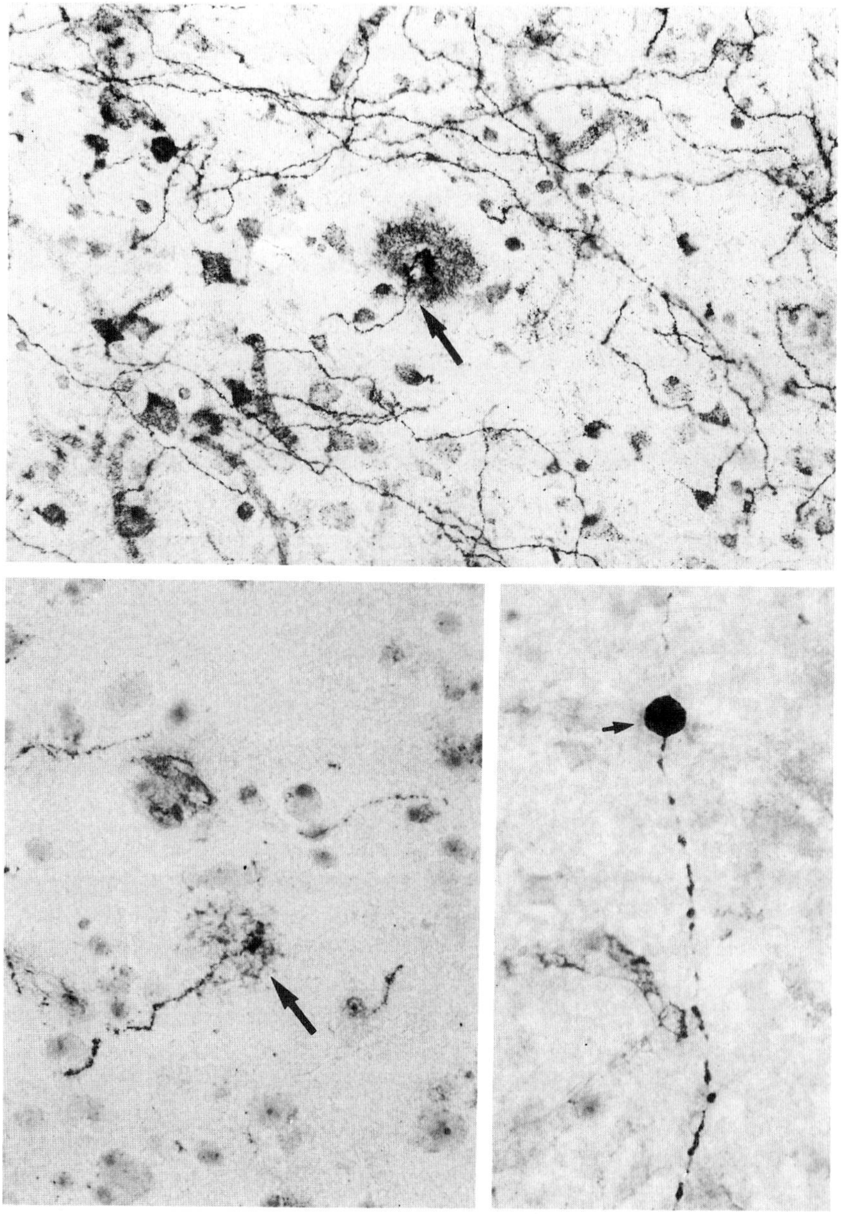

Fig. 1. Morphological alterations of cholinergic and aminergic axonal fibers in the cerebral cortex of AD cases. *a* Staining for AChE activity; arrow points to a disrupted labeled axon ending in a senile plaque; note the AChE activity in the center of the plaque.

Evidence for Damage of the Cholinergic Cell Bodies

Morphological abnormalities of the magnocellular nuclei of the forebrain have been reported by numerous investigators [1, 11, 12, 17]. These include the neurofibrillary tangles (NFT) shown by classical neuropathological studies of AD and which appear to be present in both cholinergic and noncholinergic neurons (as evaluated by AChE staining). Many recent studies have attempted to evaluate the neuronal loss in these nuclei, which varied from no loss at all to 90% cell loss [1, 11]. A similar range was found by the various investigators measuring CAT activity [11]. These discrepancies might be related in part to the anatomical complexity of the basal forebrain resulting in differences of sampling between the investigators or in the heterogeneity of the morphometrical methods applied. A comparison of their results indicates that less severe cell losses are found in studies analyzing only one level of the forebrain while analyses of serial sections usually revealed a heterogeneous pattern of cell loss, which was on the average more severe. A similar diversity of results was produced in attempts to correlate cell loss in the NBM with reductions of CAT or with the number of senile plaques in the neocortex, some authors finding significant correlations and the others not. As opposed to the numerous studies of the ventral forebrain cholinergic neurons, to the best of our knowledge, there are no studies evaluating cell loss in the caudal (pontine) cholinergic group in AD.

Evidence for a Physiopathological Role of Cholinergic Dysfunction in AD: Contribution to Clinical Symptoms

As is also shown by Mountjoy et al. in this volume, there is considerable evidence based on both biopsy and postmortem studies for an important role of the cholinergic system in the pathogenesis of clinical symptoms in AD [9, 17]. Lesions of the cholinergic system are also known to be correlated with dementia in Parkinson's disease [12], but cannot be considered as the only anatomical substrate of dementia. Indeed, there is no cholinergic deficiency in other dementing diseases such as Pick's or Huntington's disease and CAT deficiency is not constantly found in AD.

---

b Dopamine-β-hydroxylase immunostaining; swollen fiber ends in a senile plaque (↗).
c Tyrosine hydroxylase immunostaining; an axonal swelling (↗) is visible along a varicose fiber. × 350.

Noradrenergic System

Anatomy

By analogy with other mammals it can be assumed that the noradrenergic (NA) projections to the cerebral cortex in man originate primarily from the locus coeruleus (LC), a bilateral group of 12,000–18,000 melanin-containing neurons [26]. This rather small number of LC neurons contrasts with the very extensive NA fiber projections throughout the central nervous system. Therefore the NA system must be highly collateralized and have a high metabolic rate, and even limited lesions in the LC may induce considerable pathological changes in vast terminal fields. A topographic organization of the LC projections was demonstrated in rodents, but not yet in primates. Noradrenaline may not be the unique transmitter of the coeruleo-cortical pathway. Several peptides with potent neurotransmitter function – NPY, vasopressin, neurophysin – were demonstrated to be colocalized in a subpopulation of the LC neurons, thus indicating a further inhomogeneity in this nucleus.

Evidence for Damage to the LC

NFT in the LC are frequently observed in AD [16, 17]. Reductions in neuronal and nucleolar volume have also been reported suggesting a decreased protein synthesis. A gradual neuronal loss is observed in LC during normal aging reaching 25–40% at 90 years [26]. An additional neuronal loss, that ranged between 20 and 87%, was found in all AD cases investigated [20]. Marcyniuk et al. [20] localized the neuronal depopulation in the most dorsal part of the LC. This observation, if confirmed, would have important physiopathological implications because of the probable corticotropic organization of the LC. The wide variations in the damage to LC are possibly related to the age of onset of the disease: 80–87% cell losses were mainly observed in the younger patients [17]. However, such a subgrouping in terms of age does not account for all the observed differences, some being possibly related to morphometrical methods. Larger series of patients will thus have to be investigated to evaluate more precisely the interindividual variations in the LC alterations.

Evidence for an Involvement of NA Pre- and Postsynaptic Markers in the Cortical Terminal Fields

Alterations of the biochemical presynaptic markers appeared neither as extensive nor as severe as the cell loss in LC, partly for methodological

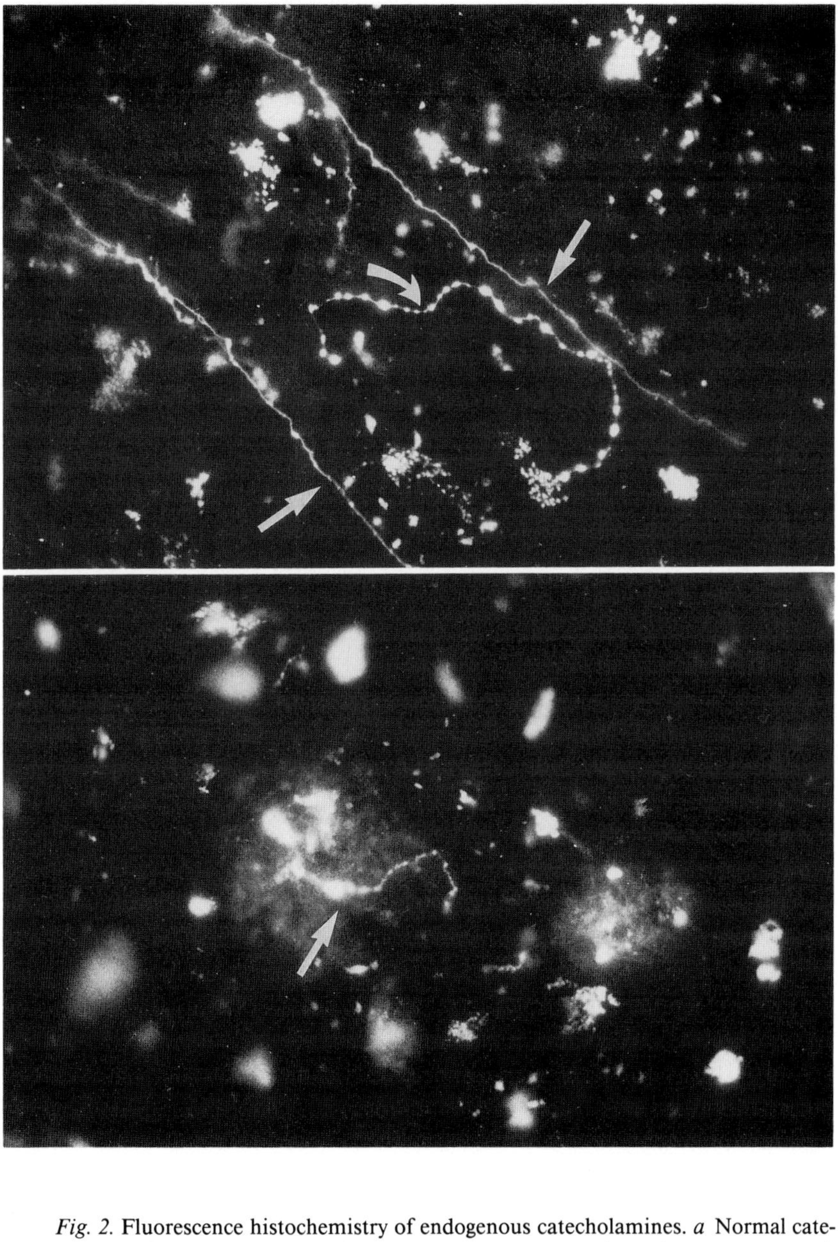

a

b

Fig. 2. Fluorescence histochemistry of endogenous catecholamines. *a* Normal catecholaminergic fibers in the human frontal cortex. A NA axon displaying numerous varicosities (curved arrow) between two dopaminergic axons (straight arrows). × 473. *b* AD: a NA-like axon with a fusiform swelling (↗) ending in a senile plaque. × 450.

reasons, such as variations in postmortem delay or short series of cases [5, 17, 22]. In neurosurgical samples of frontal and parietal cortex we observed a 90% reduction of NA levels and a striking reduction in the number of NA fibers with catecholamine fluorescence histochemistry, the remaining axons being much paler and shorter than normally [6, 7]. A characteristic feature of the AD cases was the aspect of disrupted aminergic axons with highly fluorescent round or fusiform axonal swellings (4–10 μm in size) (fig. 2). These abnormal NA axons observed within SP could reflect the participation of the degenerating NA axons to these pathological structures, known to contain modified pre- and postsynaptic elements. Such swollen aminergic terminals in SP were also identified by dopamine-β-hydroxylase and tyrosine hydroxylase immunocytochemistry (fig. 1b, c). These aspects may be interpreted as resulting from the modified axonal flow which has been postulated to be a consequence of neurofibrillary degeneration in AD.

Whether there are or not alterations of the postsynaptic NA markers is difficult to decide for the moment since the α_1-, α_2- and β-adrenergic receptors were evaluated by only two groups in the senile form of AD, with results difficult to reconcile.

Evidence for a Physiopathological Role of the NA Dysfunction in AD

Alterations of the NA system often parallel the severity of the disease and the role of the NA deficiency may be crucial since a large range of functions has been assigned to the NA system [24]. Some of these represent global functions such as the sleep-wakefulness cycle, attention eliciting situations, learning and developmental plasticity. Interestingly enough, dementia scores were correlated with damage to the LC neurons, but only in the series of AD with a relatively early onset [20]. Possibly NA dysfunction may be more closely related to some particular aspects of memory dysfunction rather than to the global intellectual ability. Indeed, NA function seems to be impaired in Korsakoff's disease [5] and a highly significant correlation was demonstrated between the extent of LC loss and the degree of memory impairment in aged rats. Dramatic improvements produced by clonidine, an α_2-agonist, in delayed response performance were observed, which might be interesting for future therapeutic attempts. To some degree the NA system exerts its role in association with the cholinergic system, probably through a common molecular mechanism. Such interactions with the cholinergic system might also occur in the human basal forebrain and NA innervation of the NBM might be reduced in AD as was suggested by Arai et al. [references in 22] showing a 77% decrease of NA in this structure.

Serotonergic System

Anatomy

The brainstem serotonergic neurons projecting to the cerebral cortex have been well defined in rodents but less so in primates. They are located in the nucleus raphe dorsalis (DRN) and centralis superior (CSN), but in the rat only one third of the DRN neurons contain 5-HT. They project in a topographically organized manner to widespread cortical targets; the rostral DRN cells innervate the frontal cortex while more caudal and medial cells project to the parietal and occipital cortex. In contrast, CSN neurons project uniformly upon the cortex and provide the hippocampal 5-HT innervation. This topography may be different in primates, since tracing studies in monkey showed that projections to the prefrontal cortex originated preferentially in the caudal raphe.

Lesions of the Main Raphe Nuclear Groups

Neuropathological changes affecting the brainstem 5-HT nuclear groups in AD have been recognized for years in presenile as well as senile forms of the disease and it was soon realized that the DRN and CSN were the most severely affected, as in some studies more than half of the remaining neurons exhibited neurofibrillary changes. Significant reductions in the nucleolar volume and the cytoplasmic RNA content were also shown, suggesting a reduced functional capacity [17]. Recently, a semiquantitative evaluation of the NFT and/or of the neuronal loss in the DRN was done in several groups of patients with different ages. As could be expected, the changes varied considerably: in the youngest patients there was a 77% decrease in the number of large neurons which are presumed to be the 5-HT ones, whereas only 8% decrease was found in the elderly patients [27]. Ichimiya et al. [14] observed an accumulation of NFT without significant neuronal loss in the CSN suggesting that this nucleus could be less impaired than the DRN in AD. More precise investigations would require the specific identification of 5-HT neurons by 5-HT or tryptophan hydroxylase (synthesizing enzyme of 5-HT) immunodetection or by the detection of its mRNA. Analysis of a larger sample of the raphe would also be requested for cell counting, since the above-mentioned works examined only one level of the DRN and this might not be completely representative of the two main nuclear groups projecting to the cerebral cortex. If the changes were not homogeneous, correlations with eventual abnormalities of the presynaptic markers would be expected to be difficult to establish.

Evidence for an Involvement of 5-HT Pre- and Postsynaptic Markers in the Cortical Terminal Fields

Presynaptic Markers. One or the other of the three presynaptic markers (endogenous 5-HT level, its main metabolite 5-HIAA or uptake capacities of (^{3}H)-5-HT) have been analyzed in 8 groups of patients and found to be affected in nearly all the cases. Similar to the other neurotransmitter systems already mentioned here, the decrease in 5-HT concentrations predominated in the cases with early onset (63–90%), while it did not exceed 53% in older patients [22].

Postsynaptic Markers. A decrease in 5-HT receptors has been detected by all investigators and concerned both receptor types: S_1 and S_2 [10]. However, the decrease in S_1 receptors was mainly found in the presenile forms whereas the decrease of S_2 was more severe (–42 to –60%) and constant in all age groups, and was considered to be more specific of AD [10]. Only few data concerning limited areas of the human cerebral cortex are available on the localization of the S_1 and S_2 postsynaptic receptors indicating a distinct laminar distribution. It is tempting to speculate that part of the 5-HT receptors might be located on the cholinergic and somatostatinergic presynaptic markers, since they are correlated quantitatively and exhibit some similarities in their cortical distribution. However, such correlation was not found by all investigators in AD, neither could it be demonstrated in demented patients with Parkinson's disease.

Evidence for a Physiopathological Role of the 5-HT Dysfunction in AD

Only a few attempts have been made to correlate the alterations of the 5-HT presynaptic markers with histopathological and clinical indices of disease severity. So far, these studies are contradictory and inconclusive. The involvement of the serotonergic system might affect the function of the NBM which receives an abundant aminergic innervation, since a 73% reduction in 5-HT levels was observed in this nucleus in AD cases [25]. At the cortical level, the impairment of the 5-HT system in AD could potentiate the NA or the cholinergic dysfunction since 5-HT probably has a facilitatory effect on both AChE and NA neurotransmission. Concerning the mechanism of 5-HT impairment, there are tenuous arguments to suggest that abnormal forms of 5-HT and 5-hydroxytryptophan are present in AD and that tryptophan treatment could improve some patients in whom an abnormal tryptophan absorption is detected.

Dopaminergic System

The cortical dopaminergic system (DA) has been the least investigated of all the aminergic systems in AD. It is usually considered to be spared since no significant pathological changes were detected in the mesencephalic DA neurons or in the DA presynaptic markers [17]. For instance, NFT were only occasionally observed in the substantia nigra, contrary to their frequent occurrence in the LC or the raphe nuclei. However, the DA innervation of the cerebral cortex is mainly provided by neurons situated in the ventral tegmental area rather than in substantia nigra and recent data in AD [19] have indicated a neuronal loss in the former but not in the latter structures. Presynaptic markers have mainly been evaluated in senile cases of AD [17, 22], where nonsignificant decreases in DA content were found in the frontal and temporal cortices. On the other hand, a large reduction of DA levels was detected in presenile AD cases (biopsy samples of the frontal cortex) [7]. Thus before concluding that the DA system totally escapes to the hypothesis of AD as an isodendritic core disease, it would be worthwhile to investigate presenile forms in which the pathological process is more severe. It would also be fruitful to investigate more specific cortical areas, especially the frontal cortex, where the DA innervation has been identified and seems to be implicated in cognitive functions [8].

Hypothalamocortical Projections

Other subcortico-cortical projections which would deserve further analysis in AD are those originating in the hypothalamus. NFT were observed in this area and severe atrophy of the magnocellular neurons (66% neuronal loss in the paraventricular and the supraoptic nucleus) was found in presenile cases which correlated significantly with the extent of LC cell loss [18]. Hypothalamocortical pathways, in non-human primates, originate in the lateral and dorsomedial part of the hypothalamus extending into the posterior and the perimammillary areas. Their cortical projection fields are remarkably widespread with the highest density in the frontal and posterior parietal cortices. The neurotransmitters of these circuits have not been yet well defined, although GABA and/or peptides (VIP, cholecystokinin, enkephalines, α-MSH-like peptide) have been implicated. The significantly lower concentration of α-MSH which has been detected in the cingulate cortex in AD patients might be related to the lesion of such a pathway.

Thalamocortical Projections

Although bilateral lesions restricted to the thalamus can occasionally be responsible for dementia, there is at present no evidence for a significant alteration of the thalamocortical tracts in AD. Pathological alterations appear to be scarce in the thalamus [16] but these nuclei have rarely been studied and in particular there have been no cell counts. The chemical nature of the thalamocortical pathways is as yet undefined. Thus further research is still requested to warrant the functional integrity of these neuronal pathways.

Conclusion

Among the numerous subcortico-cortical pathways which have been studied in AD, the available evidence indicates a frequent, if not constant, involvement of the cholinergic, noradrenergic and serotonergic systems. The severity of these lesions seems to reflect the severity of the disease process in general and is hence more pronounced in AD cases with early onset (this expression tending to be currently substituted to that of presenile dementia). However, with regards to the role of these lesions in dementia, all the attempts to establish statistical correlations between the biochemical or histopathological indices of subcortico-cortical damage and the clinical indices of mental impairment have encountered several kinds of difficulty and did not allow firm conclusions. Experimental approaches by lesions of the subcortical systems in the animal appear as yet too different from the clinical picture to derive useful information in human pathophysiology. Thus, although the ascending cholinergic projections to the hippocampus and cortex seem to play a role in memory processes, dementia is probably a much too complex clinical entity to be related to the alteration of only one neurotransmitter system.

Another unanswered question is whether the primary neurodegenerative event in AD occurs in the subcortical nuclei projecting to the cortex or in the cerebral cortex itself. This question is of interest in approaching the mechanisms of vulnerability of some neuronal subsets but makes the unverified assumption that some of the AD neuronal damage could result from retrograde or transneuronal degeneration. Recent investigations have indicated a possible selective vulnerability of the basal forebrain cholinergic neurons, since they were shown to contain a particularly high amount of the neurotrophic factor NGF [13], which is probably synthesized in the hippocampus and cortex and retrogradely transported to the basal forebrain neurons. As the forebrain cholinergic neurons appear to be particularly dependent on this factor for their survival in experimental conditions [13], new research strategies have been indicated which are being actively investigated. However, in a recent report by Atterwill and Bowen [4], no change in cholinergic neuronotrophic factor was found in brain tissue from Alzheimer's patients as compared with age-matched controls.

The hypothesis of a primary damage to the cortex and of secondary degeneration of the subcortical nuclei has been formulated by a number of authors. In particular, it has been suggested that the cortical SP could be the starting point of the neuronal degeneration, possibly induced by some neurotoxic mechanism in relationship with a breakdown of the blood-brain barrier [17]. However, the fact that SP contains abnormal neurites with many different neurochemical specificities, AChE [2], NA [6, 7], as well as peptides (SST, NPY, VIP, enk.) and GABA [see 2] and the presence of numerous NFT in both subcortical and cortical structures would rather suggest that AD results from a pluricentric involvement of several vulnerable sets of neurons in both the subcortical nuclei and the cortex.

References

More detailed references can be found in the following restrictive list of general reviews or articles.

1 Arendt, T.; Bigl, B.; Tennstedt, A.; Arendt, A.: Neuronal loss in different parts of the nucleus basalis is related to neuritic plaque formation in cortical target areas in Alzheimer's disease. Neuroscience *14:* 1–14 (1985).

2 Armstrong, D.M.; Bruce, G.; Hersh, L.B.; Terry, R.D.: Choline acetyltransferase immunoreactivity in neuritic plaques of Alzheimer brain. Neurosci. Lett. *71:* 229–234 (1986).

3 Arnsten, A.F.T.; Goldman-Rakic, P.S.: Alpha-2-adrenergic mechanisms in prefrontal cortex associated with cognitive decline in aged nonhuman primates. Science *230:* 1273–1276 (1985).

4 Atterwill, C.K.; Bowen, D.M.: Neurotrophic factor for central cholinergic neurones is present in both normal and Alzheimer brain tissue. Acta neuropath. *69:* 341–342 (1986).

5 Berger, B.: Anomalies des neurotransmetteurs dans la maladie d'Alzheimer. Revue neurol. *140:* 539–552 (1984).

6 Berger, B.; Escourolle, R.; Moyne, M.A.: Axones catécholaminergiques du cortex cérébral humain. Observation en histofluorescence, de biopsies cérébrales dont 2 cas de maladie d'Alzheimer. Revue neurol. *132:* 183–194 (1976).

7 Berger, B.; Tassin, J.P.; Rancurel, G.; Blanc, G.: Catecholaminergic innervation of the human cerebral cortex in presenile and senile dementia. Histochemical and biochemical studies; in Usdin, Sourkes, Youdim, Enzymes and neurotransmitters in mental disease, pp. 317–328 (Wiley, Chichester 1980).

8 Brozoski, T.J.; Brown, R.M.; Rosvold, H.E.; Goldman, P.S.: Cognitive deficit caused by regional depletion of dopamine in prefrontal cortex of rhesus monkey. Science *205:* 929–932 (1979).

9 Collerton, D.: Cholinergic function and intellectual decline in Alzheimer's disease. Neuroscience *19:* 1–28 (1986).

10 Cross, A.J.; Crow, T.J.; Ferrier, I.N.; Johnson, J.A.: The selectivity of the reduction of serotonin S_2 receptors in Alzheimer-type dementia. Neurobiol. Aging *7:* 3–7 (1986).

11 Etienne, P.; Robitaille, Y.; Wood, P.; Gauthier, S.; Nair, N.P.V.; Quirion, R.: Nucleus basalis neuronal loss, neuritic plaques and choline acetyltransferase activity in advanced Alzheimer's disease. Neuroscience *19:* 1279–1291 (1986).

12 Gaspar, P.; Gray, F.: Dementia and idiopathic Parkinson's disease. A neuropathological study of 32 cases. Acta neuropath. *64:* 43–52 (1984).

13 Hefti, F.; Weiner, W.J.: Nerve growth factor and Alzheimer's disease. Ann. Neurol. *20:* 275–281 (1986).

14 Ichimiya, Y.; Arai, H.; Kosaka, K.; Iizuka, R.: Morphological and biochemical changes in the cholinergic and monoaminergic systems in Alzheimer-type dementia. Acta neuropath. *70:* 112–116 (1986).

15 Kasa, P.: The cholinergic systems in brain and spinal cord. Prog. Neurobiol. *26:* 211–272 (1986).

16 Kemper, T.: Neuroanatomical and neuropathological changes in normal aging and in dementia; in Albert, Clinical neurology of aging, pp. 9–52 (Oxford University Press, London 1984).

17 Mann, D.M.A.: The neuropathology of Alzheimer's disease: a review with pathogenic aetiological and therapeutic considerations. Mech. Age. Dev. *31:* 213–255 (1985).

18 Mann, D.M.A.; Yates, P.O.; Marcyniuk, B.: Changes in Alzheimer's disease in the magnocellular neurons of the supraoptic and paraventricular nuclei of the hypothalamus and their relationship to the noradrenergic deficit. Clin. Neuropath. *4:* 127–134 (1985).

19 Mann, D.M.A.; Yates, P.O.; Marcyniuk, B.: Dopaminergic neurotransmitter systems in Alzheimer's disease and Down's syndrome at middle age. J. Neurol. Neurosurg. Psychiat. *50:* 341–344 (1987).

20 Marcyniuk, B.; Mann, D.M.A.; Yates, P.O.: The topography of cell loss from locus coeruleus in Alzheimer's disease. J. neurol. Sci. *76:* 335–345 (1986).

21 Mash, D.C.; Flynn, D.D.; Potter, L.T.: Loss of M_2 muscarine receptors in the cerebral cortex in Alzheimer's disease and experimental cholinergic denervation. Science *228:* 1115–1117 (1985).

22 Palmer, A.M.; Wilcock, G.K.; Esiri, M.M.; Francis, P.T.; Bowen, D.M.: Monoaminergic innervation of the frontal and temporal lobes in Alzheimer's disease. Brain Res. *401:* 231–238 (1987).

23 Rossor, M.N.: Parkinson's disease and Alzheimer's disease as disorders of the isodendritic core. Br. med. J. *283:* 1588–1590 (1981).

24 Sara, S.J.: The locus coeruleus and cognitive function: attempts to relate noradrenergic enhancement of signal/noise in the brain to behavior. Physiol. Psychol. *13:* 151–162 (1985).
25 Sparks, D.L.; Marksbery, W.R.; Slevin, J.T.: Alzheimer's disease: monoamines and spiperone binding reduced in nucleus basalis. Ann. Neurol. *19:* 602–604 (1986).
26 Vijayashankar, N.; Brody, H.: A quantitative study of the pigmented neurons in the nuclei coeruleus in man as related to aging. J. Neuropath. *38:* 490–497 (1979).
27 Yamamoto, T.; Hirano, A.: Nucleus raphe dorsalis in Alzheimer's disease: neurofibrillary tangles and loss of large neurons. Ann. Neurol. *17:* 573–577 (1985).

Brigitte Berger, MD, INSERM U.106, Bâtiment de Pédiatrie,
Hôpital Salpêtrière, 47, bld de l'Hôpital, F–75651 Paris Cedex 13 (France)

II. Quantitative Histological Aspects of the Aging Brain

Interdiscipl. Topics Geront., vol. 25, pp. 74–89 (Karger, Basel 1988)

Number of Plaques and Tangles, Loss of Neurons: Their Correlation with Deficient Neurotransmitter Synthesis and the Degree of Dementia

C.Q. Mountjoy

St. Andrews Hospital, Northampton, and University of Cambridge Clinical School, Cambridge, UK

Introduction

Any paper which examines the relationship between quantitative neuropathological counts and measures of the severity of dementia and the alterations of neurotransmitters must be considered in relation to the problems associated with their measurement and the statistical methods which are used to correlate them. Because most of the studies to be considered depend on correlation coefficients, it is important to remember that these coefficients are merely a reflection of the fact that two variables are moving in direct relationship with each other irrespective of whether there is a direct and causal relationship between the two variables concerned.

The majority of studies have had to rely on post-mortem examination and on relatively simple measures of dementia such as those described by the group working in Newcastle [Blessed et al., 1968]. They used two separate types of rating scales: one depended on nurses' and relatives' observations of the patient's behaviour and ability to cope with items of everyday living (Dementia Score (DS)) and the other on the patient's ability to recall information and to remember new material (Information, Memory and Concentration Test (IMCT)). These scales, though simple, are easy to administer and have a high face validity although there is some redundancy as shown by Hodkinson [1972] in the case of the IMCT. In the Newcastle studies the scales provided a clear cut-off point between functional and organic illness; a score of 6 on the DS and of 18 on the IMCT. Further evidence of the validity of the DS comes from its alteration in relation to the duration of the dementia [Mountjoy et al., 1986].

The ease of administration and the utility of these scales in measuring overall disability justifies their use, but it is impossible to obtain from them specific subsets which can be directly related to particular behaviour in order that they can be correlated with specific or localized biochemical or pathological changes.

More detailed psychological tests can be performed on patients with early dementia who are willing to undergo brain biopsy to establish a diagnosis. For many investigators this procedure was unacceptable on ethical grounds and of course it could only provide information about a small area of cortex. However, the Manchester group has produced clear evidence of the utility of biopsy in diagnosis and has shown that biopsy had no adverse effect on the subjects [Neary et al., 1986a].

Biochemical Disorders Associated with Alzheimer's Disease

The cholinergic deficit in Alzheimer's disease has been well documented since 1976 [Bowen et al., 1976; Davies and Maloney, 1976; Perry et al., 1977] and has been a consistent finding in all subsequent studies. Deficits in other neurotransmitters have been demonstrated. Concentrations of noradrenaline (NA) and its metabolite 3-methoxy-4-hydroxyphenylglycol (MHPG) have been found to be reduced in the cerebral cortex at post-mortem examination. Similarly the activity of dopamine beta-hydroxylase (DBH), the specific biosynthetic marker of noradrenergic neurons, is reduced in the cerebral cortex [Cross et al., 1981; Rossor and Iversen, 1986].

Reductions in the 5-hydroxytryptamine (5-HT) system have been reported by Adolfsson et al. [1979] in post-mortem tissue, and by Benton et al. [1982] in neocortical biopsy specimens. Subsequently, Bowen et al. [1983] showed a reduction in 5-HT_1 receptors in SDAT, and Reynolds et al. [1984] showed significant loss of ketanserin binding, indicating loss of 5-HT_2 receptors. They were unable to show significant differences between dements and controls in respect of 5-HT and 5-hydroxyindoleacetate (5-HIAA) though there was a trend for the levels in dements to be lower than in controls. Gamma-aminobutyric acid (GABA) is not generally reduced but there is some evidence that the level of the GABA is reduced in the temporal lobes of patients dying before the age of 79 [Rossor et al., 1982]. There appears to be no alteration in the levels of dopamine in the cortex. The only neuropeptide consistently found to be significantly reduced in dementia is somatostatin [Rossor et al., 1980; Davies et al., 1980].

Neuropathological Changes Associated with Alzheimer's Disease

The classical microscopical changes associated with a diagnosis of Alzheimer's disease are the senile plaque [Blocq and Marinesco, 1892], the neurofibrillary tangle [Alzheimer, 1907a, b] and neuronal loss. Each of these lesions will be considered in relation to the severity of dementia and the biochemical measures in dementia.

Plaques

Correlation of Plaque Counts with Severity of Dementia

Traditionally the plaques and tangles have been demonstrated by silver stains or by stains for amyloid such as thioflavine-S. Although different stains produce different absolute numbers of plaques in adjacent sections, the counts correlate highly with each other.

The Newcastle group who were pioneers in the quantitative investigation of plaques in relation to the severity of dementia favour the use of a modification of the von Braunmühl method. With this method the plaque appears as a discrete area of dark silver impregnation. Tomlinson and colleagues developed a method of counting which had a high inter-rater reliability ($r = +0.99$) between a trained neuropathologist on the one hand and two psychiatrists without previous experience of neuropathological techniques on the other [Blessed et al., 1968].

Slides taken from twelve samples of cortex were examined to find an area of each slide which was representative of the general density of plaques. Counts were then made by the use of a graticule in five fields in each of the sections and an average count for each person was derived by adding the total number of plaques and dividing by the number of fields. Using this method, Blessed et al. [1968] showed an orderly relationship between plaque counts in the cortex and the severity of the dementia as measured by the DS and IMCT. The correlations with dementia were high (DS: r = +0.77; IMCT: r = −0.59). However, there was considerable scatter in the number of plaques, so that some patients showed a marked impairment on the DS yet had a low plaque count, while others scored well within the range for normals but had large plaque counts. Nonetheless, the correlation coefficients obtained were high and this suggested that plaque formation was directly related to the dementing process and could no longer be regarded as a mere epiphenomenon.

Wilcock et al. [1982] counted plaques in the frontal, temporal, parietal and occipital lobes. They also studied several areas of the hippocampus, the H_1, H_2 and H_3 areas, the subiculum and the parahippocampus. They correlated the plaque counts in each of these areas with a four-point DS based on information obtained from relatives and the patient's performance on a mental test score [Hodkinson, 1972] and/or modified Kew test described by Hare [1978]. The correlations obtained by these authors were lower than those obtained by Blessed et al. [1968] and they were not statistically significant in the frontal or temporal lobes, or in the subiculum or the H_2 sector of the hippocampus. In the other areas they were statistically significant though the correlation was relatively small and the highest correlation was found in the occipital lobe (r = 0.43).

The studies of Wilcock et al. [1982] and those of Blessed et al. [1968] are not directly comparable because the rating instruments used to assess the degree of dementia were different and the method of relating these scores to pathological measures was different. The Newcastle study averaged counts for the whole brain whereas Wilcock et al. [1982] correlated pathological changes in each lobe separately with the DS.

A later study [Mountjoy et al., 1986] correlated the DS of Blessed et al. [1968] with plaque counts and estimates of tangle formation in nine cortical areas. Plaque counts correlated highly and significantly in all areas with the DS. Neary et al. [1986b] correlated plaque counts in biopsy material from the right middle temporal gyrus with a clinical rating scale, the Wechsler Adult Intelligence Scale [Wechsler, 1944], a modified version of the Token Test [De Renzi and Vignolo, 1962] and Visual Reaction Time Task. The only significant correlation was between plaque count and the verbal subscales of the WAIS (r = −0.45). This finding cannot be explained by low plaque counts in early cases in whom the diagnosis is in doubt because the values for the individual cases show the plaques to be present in moderate numbers.

Correlations of Plaques with Biochemical Lesions

Perry et al. [1978] demonstrated a highly significant correlation between cortical choline acetyltransferase (ChAT) activity and mean plaque count in 51 cases who had been diagnosed normal or as suffering from depression, multi-infarct dementia or SDAT. The correlations between ChAT activity and plaque formation were high ($r = -0.82$, $p < 0.001$) for the total group, but no data were presented for the demented group alone. Acetylcholinesterase (AChE) activity was significantly correlated with plaque counts but muscarinic cholinergic receptor binding was not. Subsequently, Perry et al. [1981a] demonstrated significant correlations between plaque counts and ChAT in nondemented and demented cases combined, and for Alzheimer cases alone in the temporal lobe ($r = -0.65$).

By contrast, Wilcock et al. [1982], using an average plaque count for frontal and temporal lobe separately, were unable to show a significant correlation between the ChAT activity and plaque count in frontal ($r = -0.21$) or temporal ($r = -0.24$) lobes. Mountjoy et al. [1984] studied 25 demented cases and 25 controls and demonstrated significant negative correlations between ChAT activity and plaque counts in the nine areas of cortex studied. The correlations for the total group were similar to those reported by Perry et al. [1978] and were of the order of $r = 0.63$. However, in the demented group alone, correlations were lower and of the order of $r = 0.45$.

Perry et al. [1980] demonstrated that some plaques and neurofibrillary tangles in the hippocampus in normal elderly subjects and patients suffering from senile dementia stained positively for AChE. The positive staining indicated that senile plaques in the hippocampus were at least partially derived from cholinergic fibres. In the noradrenergic systems, Perry et al. [1981] found no significant correlations between plaque counts and DBH activity in the cerebral cortex. No significant correlations were found between plaque counts and GAD activity [Perry et al., 1978] or GABA activity [Mountjoy et al., 1984]. Perry et al. [1981a] showed a correlation between substance P and cholecystokinin and plaque counts in a combined group of demented and nondemented subjects, but significant correlations were not found in the demented group alone and there is no clear evidence that these peptides are reduced in dementia.

Somatostatin has been shown to correlate significantly with plaque counts in Brodmann area 21 [Dawbarn et al., 1986]. Somatostatin has been demonstrated in neuritic plaques and is thought to represent degeneration of the somatostatin-containing neurites, [Morrison et al., 1985]. More recently, Struble et al. [1986], using immunocytochemical methods, have shown that the neurites in plaques are derived from multiple transmitter systems, including some not previously identified to be at risk in Alzheimer's disease.

Neurofibrillary Tangles

Correlations of Tangles with Severity of Dementia

Because the tangles are smaller than plaques and are less evenly distributed than the plaques, it is hard to obtain a reliable figure for tangle counts unless the normal neurons are also counted. Such work is extremely tedious and few have attempted it. Some authors have found it satisfactory to estimate the density of tangles on a four-point scale [Corsellis, 1962; Dayan, 1970; Mountjoy et al., 1983]. Among the few who have counted the number of tangles are Wilcock et al. [1982]. These authors were able to make tangle counts in the same areas of cortex as the plaque counts already discussed [Wilcock et al., 1982]. The tangles were significantly correlated with the DS in all areas examined except the subiculum and H_2 sector of the hippocampus. The highest correlation coefficient was $r = 0.59$ in the temporal lobe. Because the correlation coefficients between the DS and the tangles were higher than for plaques, the authors concluded that the tangle was of more importance in the dementing process than the plaque.

Mountjoy et al. [1986] correlated the DS of Blessed et al. [1968] with estimates of neurofibrillary tangles rated on a four-point scale of none, mild, moderate and severe. This method lacks the rigour of that employed by Wilcock et al. [1982] though it does overcome the difficulty of choosing a representative sample of cortex for the count of tangles. Like Wilcock et al. [1982], the level of the correlation coefficient for the tangles was higher than that for the plaques.

It should be realized that there is a problem in making any correlation of tangles with anything else if the population studied includes both controls and dements. The reason for this is that although there is a steady scatter of scores between controls and dements in respect of psychological measures such as the DS, there is a dichotomy in the distribution of the tangles. To a large extent the pathological diagnosis of SDAT is made on the basis of neocortical tangles, and many pathologists feel that their presence is a necessary condition to make the diagnosis. Therefore, the controls will have very few or no neocortical tangles and dements will have many. The result is that any correlation coefficient will be high because of a statistical 'dumbell' effect. Neary et al. [1986b] in their biopsy series found that tangles correlated significantly with the verbal subscale of the WAIS ($r = -0.47$; $p < 0.05$) and the Token Test ($r = -0.65$).

Correlations of Tangles with Biochemical Lesions

Wilcock et al. [1982] in a combined group of controls and dements found significant correlations between ChAT activity and tangle counts in the frontal ($r = -0.3$) and temporal lobe ($r = -0.58$). When the demented patients alone were examined, the correlation remained significant only in the temporal lobe ($r = -0.43$). Mountjoy et al. [1984] showed significant correlations between ChAT activity and estimates of neurofibrillary change based on a 0–3 scale. These changes were only significant in the superior, middle and inferior frontal and middle temporal gyri and were of the order of $r = 0.5$. Correlations for combined demented and control groups produced very large correlation coefficients of the order of $r = 0.8$ for the reasons already outlined. There was a single significant correlation between GABA and tangles in the superior temporal gyrus ($r = -0.53$; $p = 0.006$) but this may have been no more than a chance finding in view of the large number of correlations calculated.

Neurons

Correlations of Neuronal Counts with Severity of Dementia

Attempts have been made to count cortical neurons by means of image-analysing equipment. The results of these investigations have been conflicting in that an early study by Tomlinson and Henderson [1976] failed to demonstrate any significant loss of cortical neurons. This appeared to be confirmed by Terry et al. [1977], but subsequently, Terry et al. [1981] demonstrated a significant reduction in the number of neurons in demented patients compared with controls in the middle frontal region and superior temporal gyrus. Mountjoy et al. [1983] also demonstrated a significant reduction in cortical neuronal counts in the frontal and temporal lobes of demented patients.

Mountjoy et al. [1986] sought correlations between the DS and cortical neuronal counts made with the aid of an image-analysing computer.

Neuronal counts made in this way correlated weakly and poorly with the DS and the only statistically significant correlation found in the nine cortical areas examined was in the anterior cingulate gyrus ($r = -0.5$; $p < 0.005$). The poor correlation between neuronal counts and DS may reflect problems associated with the method of making the neuronal counts or be the consequence of biochemical lesions preceding neuronal death, and being of greater importance in aetiology of dementia than neuronal death.

By contrast, Neary et al. [1986b] who studied biopsy material from the right middle temporal gyrus in 17 cases of Alzheimer's disease found that a statistically significant correlation existed between a number of psychological tests and cell loss in layers III and V, reductions in nuclear and nucleolar volume and cytoplasmic RNA content. This study has the advantage of using biopsy rather than post-mortem tissue and this, together with the high intercorrelations between the neuronal measures, may explain the differences between this study and that of Mountjoy et al. [1986]. That the more sophisticated psychological tests possible in a biopsy series correlate better with neuronal measures than plaque counts in relatively early cases points to the primacy of neuronal lesions.

Correlations of Neurons with Biochemical Measures

Cholinergic input to the cortex comes mainly from the substantia innominata [Emson and Lindvall, 1986]. Evidence of neuronal loss was first described by Pilleri [1966] and has subsequently been confirmed by him and by other workers. Perry et al. [1980], using an AChE stain, showed a 33% reduction in neuron density in a group of 36 Alzheimer's disase and SDAT brains. This was a relatively low reduction compared to the reported ChAT reduction of 75–90%, and led these authors to suggest that the loss of cells in the basal nucleus of Meynert could be secondary to cortical changes.

Pearson et al. [1983] found no significant reduction in cells stained for ChAT or AChE in the basal nucleus of Meynert, though the size of the cells was significantly reduced in SDAT patients compared with controls. Loss of ChAT activity has been shown in the substantia innominata [Rossor et al., 1982]. Wilcock et al. [1983] compared 6 patients with SDAT with 4 normals and showed a 50% reduction in basal nucleus count which was associated with a similar percentage reduction in the average ChAT activity. Correlations between count and cortical cholinergic activity are not presented in the paper, but from the raw data it is possible to calculate the correlation coefficient between basal nucleus count and frontal and temporal ChAT activity (r = 0.23 and r = 0.08, respectively).

Correlations between counts of cortical neurons and ChAT activity in the corresponding area in the opposite hemisphere showed that there were significant positive correlations between counts and ChAT activity in the superior, middle and inferior frontal gyri and the superior and middle temporal gyri in a combined group of demented and nondemented patients. No significant correlations were found when the demented or control groups were examined separately [Mountjoy et al., 1984].

Recently, Perry et al. [1984] have been able to examine the intralaminar distributions of transmitter and nontransmitter enzyme activities in the midtemporal cortex of normal individuals and in established cases of Alzheimer's disease. They were able to show that ChAT and AChE activities were relatively higher in the outer cortical layers, particularly in the case of ChAT, in layers II and IV. Activities in these layers and in all other layers of grey matter were reduced but the white matter was not affected. They found no differences in the distribution of pseudocholinesterase.

Abnormalities in NA have been found in SDAT by a number of authors. Counts of cells in the locus ceruleus, the nucleus which provides the noradrenergic input to the cortex, have been shown to be reduced by 50–60% in SDAT [Forno, 1978; Bondareff et al., 1982; Tomlinson et al., 1976; Mann et al., 1982, 1984; Iversen et al., 1983]. In general the greatest loss has been found in those patients dying at a relatively young age [Bondareff et al., 1982; Perry et al., 1981b; Mann et al., 1982, 1984; Iversen et al., 1983], but large losses are by no means confined to the younger age group, as shown by Iversen et al. [1983]. Perry et al. [1981b] found no significant correlation between locus ceruleus counts and cortical DBH activity in Brodmann areas 10 and 21, in either normals or patients suffering from SDAT. Iversen et al. [1983], however, reported a 60% reduction of NA activity in Brodmann areas 21 and 38, which was comparable to the percentage reduction in locus ceruleus count when compared with normals. In this study, by contrast, ChAT was reduced by only 40%, and there was no change in the amount of dopamine in the cortex.

The activity of the GABAergic marker enzyme glutamic acid decarboxylase (GAD) is reduced in some areas of post-mortem cerebral cortex [Bowen et al., 1976] and in the midbrain in cases of SDAT. However, this reduction is probably due to agonal effects because biopsy tissue has normal GAD activity in the temporal cortex in Alzheimer's disease. GABA concentrations which appear to be unaffected by the agonal state have been shown to be reduced in the temporal lobe in SDAT by Rossor et al. [1982]. Perry et al. [1984] found a much more extensive intralaminar variation in GAD activity compared with the level of GABA. They found neither the levels of GAD activity and GABA nor the intralaminar patterns of GABA to be altered in Alzheimer's disease. Mountjoy et al. [1984] found that correlations between GABA activity and neuronal counts were only significant in two of the nine areas studied (inferior frontal and superior temporal gyri) in the combined group of dements and controls.

Mann and Yates [1983] showed that there was a significant reduction in nucleolar volume and RNA content in the cells of both the medial and lateral dorsal tegmental nuclei, the cells of origin of some of the 5-HT input to the cortex. Direct correlations between these counts and presence of 5-HT were not reported, but there is enough evidence to suggest that the changes found in 5-HT and the dorsal tegmental nucleus are similar to those found in respect of the locus ceruleus and NA and between cell counts of the basal nucleus of Meynert and ChAT activity.

A number of neuropeptides have been studied in SDAT. Three of these, vasopressin [Rossor et al., 1980], cholecystokinin [Perry et al., 1981a] and vasoactive intestinal peptide [Perry et al., 1981a; Rossor et al., 1980], do not differ significantly from normal, but the level of substance P has been reported by Davies et al. [1982] to be reduced by 30–50%. Preliminary evidence [Perry et al., 1983] on the distribution of somatostatin-like immunoreactivity in temporal cortex suggests that reductions in activity of this neuropeptide in Alzheimer's disease may be most marked in the inner cortical layers. Roberts et al. [1985] showed in Alzheimer's disease that the somatostatin-immunoreactive perikaryon exhibited changes consistent with neuronal degeneration.

Correlations of Biochemical Lesions with Severity of Dementia

The first effort to relate biochemical activity post-mortem with the severity of dementia measured in life was by Perry et al. [1978]. They showed that the IMCT [Blessed et al., 1968] correlated highly and significantly with the measurement of ChAT, the biosynthetic enzyme for the cholinergic system ($r = 0.81$). By contrast, the activity of GAD was not significantly correlated with the IMCT ($r = 0.16$).

In a later paper, Perry et al. [1981a] replicated their findings concerning the activity of both ChAT and GAD. It also showed low and statistically nonsignificant correlation coefficients between IMCT and DBH, cholecystokinin, vasoactive intestinal peptide, substance P and somatostatin. The finding regarding DBH was confirmed in a later publication [Perry et al., 1981b]. Wilcock et al. [1982] found a significant correlation between their four-point DS and ChAT activity. The ChAT activity was most reduced in the temporal lobe and the correlation of the DS here was better ($r = -0.67$) than in the frontal lobe ($r = -0.38$).

Gottfries et al. [1983] measured 5-HT, 5-HIAA, NA, MHPG, DA, HVA, ChAT, MAO-A and MAO-B in the hypothalamus, caudate nucleus, hippocampus and cingulate gyrus and correlated each of these measurements with a DS completed by the head nurse the day after the death of the patient. These authors found that the most severely demented patients had positive correlations between NA and MHPG, in other words the most demented had the highest levels of the substances. They found ChAT activity in the caudate to be negatively but not quite significantly correlated with mental impairment ($r = 0.55$). There were no significant correlations in respect of the measurements.

In a later study, Mountjoy et al. [1986] studied a group of 30 patients in whom ChAT and GABA activity had been measured in 16 areas of the cortex. ChAT activity correlated negatively and highly significantly with DS in all 16 areas of cortex, whereas GABA had a single significant correlation in Brodmann area 9, which was just significant at the 5% level of probability. Although the DS correlated significantly with ChAT activity in all areas of the cerebral cortex, it did not correlate significantly with the changes in ChAT activity in the cerebellar cortex. This is especially interesting because it demonstrates that it is changes in ChAT activity in the cerebral cortex alone which relates to the measure of intellectual decline.

In the same paper a more extensive series of patients and controls varying in number between 60 and 74 was studied in Brodmann areas 10, 21 and 24, in respect of ChAT and GABA activity, NA, DA and somatostatin. Once again, the ChAT activity was correlated highly with the DS in each of these areas but GABA was not. NA was significantly correlated in only Brodmann area 24, where the level of correlation was of the same order as that found for ChAT. In Brodmann area 21, somatostatin and ChAT activity were equally well correlated with the DS. A peculiar aspect of this study was a positive correlation between dopamine activity and the DS – in other words, the higher the dopamine measurement in Brodmann area 21, the more demented the subject. The fact that the most consistent abnormality found in dementia is lowered ChAT activity and that this activity correlates most highly with measures of severity of dementia suggests to some that an intervention in the cholinergic system should be the most promising therapeutic intervention. However, early clinical trials carefully carried out have failed to show any significant treatment effect [Hollander et al., 1986]. The rational treatment of dementia can only take into account the wide but selective deficits that have been described. A proper evaluation of the problems associated with dementia demands a prospective study of representative cohorts of elderly subjects.

Conclusion

Comparison of the papers summarized in this chapter is almost impossible because of differences in the rating scales used and the methods employed to measure the pathological and biochemical variables in different areas. Although there is inevitably some 'noise' in the results, the post-mortem studies do have a number of features in common. There are usually high correlations between the measures of dementia and the senile plaques and neurofibrillary tangles. This is reassuring in that it is these pathological changes which have been regarded as central to the pathological diagnosis of Alzheimer's disease. The high correlations between dementia measures and pathological change indicate that these changes make a quantitative contribution to the dementing process and not just a qualitative one. The higher correlations between dementia measures and tangles compared with plaque counts suggests that the tangle is probably of greater importance than the plaque in the dementing process. Again, this is necessary in view of the fact that neocortical tangles have been regarded as central to the dementing process. As would be expected, those biochemical deficits which have been found in dementia and to correlate with its severity also correlate highly with those pathological lesions which also correlate highly with severity of dementia measured in the same way.

These apparently clear findings concerning the plaque and the tangle have been called in question by the biopsy series of Neary et al. [1986a, b] where more sophisticated psychological tests failed to correlate closely with plaque or tangle counts though they did with neuronal measures which had not provided good correlations in post-mortem studies. These findings are important and point to the need to elucidate the nature of neuronal changes which may be the prelude to tangle formation in Alzheimer's disease.

References

Adolfsson, R.; Gottfries, C.G.; Roos, B.E.; Winblad, B.: Changes in brain catecholamines in patients with dementia of Alzheimer's type. Br. J. Psychiat. *135:* 216–223 (1979).

Alzheimer, A.: Über eine eigenartige Erkrankung der Hirnrinde. Allg. Z. Psychiat. *64:* 146–148 (1907a).

Alzheimer, A.: Über eine eigenartige Erkrankung der Hirnrinde. Zentbl. ges. Neurol. Psychiat. *18:* 177–179 (1907b).

Benton, J.S.; Bowen, D.M.; Allen, S.J., et al.: Alzheimer's disease as a disorder of isodentritic core. Lancet *i:* 456 (1982).

Blessed, G.; Tomlinson, B.E.; Roth, M.: The association between quantitative measures of dementia and of senile change in the cerebral grey matter of elderly subjects. Br. J. Psychiat. *114:* 797–811 (1968).

Blocq, P.; Marinesco, G.: Sur les lésions et la pathogénie de l'épilepsie dite essentielle. Sem. méd. *12:* 445–446 (1892).

Bondareff, W.; Mountjoy, C.Q.; Roth, M.: Loss of neurons of origin of the adrenergic projection to the cerebral cortex (nucleus locus coeruleus) in senile dementia. Neurology *32:* 164–168 (1982).

Bowen, D.M.; Allen, S.J.; Benton, J.S.; Goodhardt, M.J.; Haan, E.A.; Palmer, A.M.; Sims, N.R.; Smith, C.C.T.; Spillane, J.A.; Esiri, M.M.; Neary, D.; Snowden, J.S.; Wilcock, G.K.; Davison, N.: Biochemical assessment of serotonergic and cholinergic dysfunction and cerebral atrophy in Alzheimer's disease. J. Biochem. *41:* 266–272 (1983).

Bowen, D.M.; Smith, C.B.; White, P.; Davison, A.N.: Neurotransmitter-related enzymes and indices of hypoxia in senile dementia and other abiotrophies. Brain *99:* 459–496 (1976).

Corsellis, J.A.N.: Mental illness and the ageing brain (Oxford University Press, London 1962).

Cross, A.J.; Crow, T.J.; Johnson, J.A., et al.: Monoamine metabolism in senile dementia of Alzheimer type. J. neurol. Sci. *60:* 383–392 (1981).

Davies, P.; Katz, D.A.; Crystal, A.H.: Choline acetyltransferase, somatostatin and substance P in selected cases of Alzheimer's disease; in Corkin, Davis, Growdon, Usdin, Wurtman, Alzheimer's disease: a report of progress, vol. 19: Ageing, pp. 9–14 (Raven Press, New York 1982).

Davies, P.; Katzman, R.; Terry, R.D.: Reduced somatostatin-like immunoreactivity in cerebral cortex from cases of Alzheimer senile dementia. Nature, Lond. *288:* 279–280 (1980).

Davies, P.; Maloney, A.J.F.: Selective loss of central cholinergic neurons in Alzheimer's disease. Lancet *ii:* 1403 (1976).

Dawbarn, D.; Rossor, M.N.; Mountjoy, C.Q.; Roth, M.; Emson, P.C.: Decreased somatostatin immunoreactivity but not neuropeptide Y immunoreactivity in cerebral cortex in senile dementia of Alzheimer type. Neurosci. Lett. *70:* 154–159 (1986).

Dayan, A.D.: Quantitative histological studies on the aged human brain. II. Senile plaques and neurofibrillary tangles in senile dementia (with an appendix on their occurrence in cases of carcinoma). Acta neuropath. *16:* 95–102 (1970).

De Renzi, E.; Vignolo, L.A.: The Token Test: A sensitive test to detect receptive disturbances in aphasics. Brain *85:* 665–678 (1962).

Emson, P.C.; Lindvall, O.: Neuroanatomical aspects of neurotransmitters affected in Alzheimer's disease. Br. med. Bull. *42:* 57–62 (1986).
Forno, L.S.: The locus coeruleus in Alzheimer's disease. J. Neuropath. exp. Neurol. *37:* 614 (1978).
Gottfries, C.G.; Adolfsson, R.; Aquilonius, S.M.; Carlsson, A.; Eckernas, S.A.; Nordberg, A.; Oreland, L.; Svennerholm, L.; Wiberg, A.; Winblad, B.: Biochemical changes in dementia disorders of Alzheimer type (AD/SDAT). Neurobiol. Ageing *4:* 261–271 (1983).
Hare, M.: Clinical check-list for diagnosis of dementia. Br. med. J. *ii:* 266–267 (1978).
Hodkinson, H.M.: Evaluation of a mental test score for assessment of mental impairment in the elderly. Age Ageing *1:* 233–238 (1972).
Hollander, E.; Mohs, R.C.; Davis, K.L.: Cholinergic approaches to the treatment of Alzheimer's disease. Br. med. Bull. *42:* 97–100 (1986).
Iversen, L.L.; Rossor, M.N.; Reynolds, G.P., et al.: Loss of pigmented dopamine-β-hydroxylase positive cells from locus coeruleus in senile dementia of Alzheimer's type. Neurosci. Lett. *39:* 95–100 (1983).
Mann, D.M.A.; Yates, P.O.: Serotonin nerve cells in Alzheimer's disease. J. Neurol. Neurosurg. Psychiat. *46:* 96 (1983).
Mann, D.M.A.; Yates, P.O.; Hawkes, J.: The noradrenergic system in Alzheimer and multi-infarct dementias. J. Neurol. Neurosurg. Psychiat. *45:* 113–119 (1982).
Mann, D.M.A.; Yates, P.O.; Marcyniuk, B.: Alzheimer's presenile dementia, senile dementia of Alzheimer type and Down's syndrome in middle age form an age-related continuum of pathological changes. Neuropath. appl. Neurobiol. *10:* 185–207 (1984).
Mann, D.M.A.; Yates, P.O.; Marcyniuk, B.: Some metamorphic observations on the cerebral cortex and hippocampus in presenile Alzheimer type and Down's syndrome in middle age. J. neurol. Sci. *69:* 139–159 (1985).
Morrison, J.H.; Rogers, J.; Scherr, S.; Benoit, R.; Bloom, F.E.: Somatostatin immunoreactivity in neuritic plaques of Alzheimer's patients. Nature, Lond. *314:* 90–95 (1985).
Mountjoy, C.Q.; Rossor, M.N.; Evans, N.J.R.; Reynolds, G.P.; Evans, H.; Roth, M.; Iversen, L.L.; Emson, P.: Biochemical and neuropathological changes in the brain and their correlation to the severity of dementia in Alzheimer's disease. Biol. Psychiat. *21:* 1415–1417 (1986).
Mountjoy, C.Q.; Rossor, M.N.; Iversen, L.L.; Roth, M.: Correlation of cortical cholinergic and GABA deficits with quantitative neuropathological findings in senile dementia. Brain *107:* 507–518 (1984).
Mountjoy, C.Q.; Roth, M.; Evans, N.J.R.; Evans, H.: Cortical neuronal counts in normal elderly controls and demented patients. Neurobiol. Ageing *4:* 1–11 (1983).
Neary, D.; Snowden, J.S.; Bowen, D.M.; Sims, N.R.; Mann, D.M.A.; Yates, P.O.; Davison, A.N.: Cerebral biopsy in the investigation of presenile dementia due to cerebral atrophy. J. Neurol. Neurosurg. Psychiat. *49:* 157–162 (1986a).
Neary, D.; Snowden, J.S.; Mann, D.M.A.; Bowen, D.M.; Sims, N.R.; Nathen, B.; Yates, P.O.; Davison, A.N.: ALzheimer's disease: a correlative study. J. Neurol. Neurosurg. Psychiat. *49:* 229–237 (1986b).
Pearson, R.C.A.; Sofroniew, M.V.; Cuello, A.C.; Powell, T.P.; Eckenstein, F.; Esiri, M.M.; Wilcock, G.K.: Persistence of cholinergic neurons in the basal nucleus in a brain with senile dementia of the Alzheimer's type demonstrated by immunohistochemical staining for choline acetyltransferase. Brain Res. *289:* 375–379 (1983).

Perry, E.K.; Blessed, G.; Tomlinson, N.E.; Perry, R.H.; Crow, T.J.; Cross, A.J.; Dockray, G.J.; Dimaline, R.; Arregui, A.: Neurochemical activities in human temporal lobe related to ageing and Alzheimer-type changes. Neurobiol. Ageing *2:* 251–256 (1981a).

Perry, E.K.; Curtis, M.; Dick, D.J.; Candy, J.M.; Atack, J.R.; Bloxham, C.A.: Blessed, G.; Fairbairn, A.; Tomlinson, B.E.; Perry, R.H.: Cholinergic correlates of cognitive impairment in Parkinson's disease. J. Neurol. Neurosurg. Psychiat. *48:* 413–421 (1985).

Perry, E.K.; Perry, R.H.; Blessed, G.; Tomlinson, B.E.: Neurotransmitter enzyme abnormalites in senile dementia – choline acetyltransferase and glutamic acid decarboxylase in necropsy brain tissue. J. neurol. Sci. *34:* 247–265 (1977).

Perry, E.K.; Tomlinson, B.E.; Blessed, G.; Bergmann, K.; Gibson, P.H.; Perry, R.H.: Correlation of cholinergic abnormalities with senile plaques and mental test scores in senile dementia. Br. med. J. *ii:* 1457–1459 (1978).

Perry, E.K.; Tomlinson, B.E.; Blessed, G.; Perry, R.H.; Cross, A.J.; Crow, T.J.: Neuropathological and biochemical observations on the noradrenergic system in Alzheimer's disease. J. neurol. Sci. *51:* 279–287 (1981b).

Perry, R.H.; Blessed, G.; Perry, E.K.; Tomlinson, B.E.: Histochemical observations on cholinesterase activities in the brains of elderly normal and demented (Alzheimer-type) patients. Age Ageing *9:* 9–16 (1980).

Perry, R.H.; Candy, J.M.; Perry, E.K.: Some observations and speculations concerning the cholinergic system and neuropeptides in Alzheimer's disease; in Terry, Katzman, Biological aspects of Alzheimer's disease. Banbury Report *15:* 351–361 (1983).

Pilleri, G.: Kluver-Bucy syndrome in man. A clinicopathological contribution to the function of the medial temporal lobe structures. Psychiatria Neurol. *152:* 65–103 (1966).

Reynolds, G.P.; Arnold, L.; Rossor, M.N.; Iversen, L.L.; Mountjoy, C.Q.; Roth, M.: Reduced binding of (^{3}H)-ketanserin to cortical 5-HT_2 receptors in senile dementia of the Alzheimer type. Neurosci. Lett. *44:* 47–51 (1984).

Roberts, G.W.; Crow, T.J.; Polak, J.M.: Location of neuronal tangles in somostatin neurones in Alzheimer's disease. Nature, Lond. *314:* 92–94 (1985).

Rossor, M.N.; Emson, P.C.; Mountjoy, C.Q.; Roth, M.; Iversen, L.L.: Reduced amounts of immunoreactive somatostatin in the temporal cortex in senile dementia of Alzheimer type. Neurosci. Lett. *20:* 373–377 (1980).

Rossor, M.N.; Garret, N.J.; Johnson, A.L.; Mountjoy, C.Q.; Roth, M.; Iversen, L.L.: A post-mortem study of the cholinergic and GABA systems in senile dementia. Brain *105:* 313–330 (1982).

Rossor, M.N.; Iversen, L.L.: Non-cholinergic neurotransmitter abnormalities in Alzheimer's disease. Br. med. Bull. *42:* 70–74 (1986).

Rossor, M.N.; Iversen, L.L.; Reynolds, G.P.; Mountjoy, C.Q.; Roth, M.: Neurochemical characteristics of early and late onset types of Alzheimer's disease. Br. med. J. *288:* 361–364 (1984).

Struble, R.G.; Powers, R.E.; Kitt, C.A.; Walker, L.C.; Cook, L.C.; Price, D.L.: Transmitter specificity of neuritis in senile plaque in ageing and Alzheimer's disease. Senile dementias: early detection, pp. 230–236 (Libby Eurotext/London Press, London 1986).

Tagliavini, F.; Pilleri, G.: Neuronal counts in the basal nucleus of Meynert in Alzheimer disease and in simple senile dementia. Lancet *ii:* 469–470 (1983).

Terry, R.D.; Fitzgerald, C.; Peck, A.; Millner, J.; Farmer, P.: Cortical cell counts in senile dementia. J. Neuropath. exp. Neurol. *36:* 633 (1977).

Terry, A.D.; Peck, A.; DeTeresa, R.; Schecter, R.: Some morphometric aspects of the brain in senile dementia of the Alzheimer type. Ann. Neurol. *10:* 184–192 (1981).

Tomlinson, B.E.; Henderson, G.: Some quantitative cerebral findings in normal and demented old people; in Terry, Gerhson, Neurobiology of ageing, vol. 3, pp. 183–204 (Raven Press, New York 1976).

Wechsler, D.: The measurement of human intelligence (Williams & Wilkins, Baltimore 1944).

Wilcock, G.K.; Esiri, M.M.; Bowen, D.M.; Smith, C.T.: Correlation of cortical choline acetyltransferase activity with the severity of dementia and histological abnormalities. J. neurol. Sci. *57:* 407–417 (1982).

Wilcock, G.K.; Esiri, M.M.; Bowen, D.M.; Smith, C.C.T.: The nucleus basalis in Alzheimer's disease: cell counts and cortical biochemistry. Neuropath. appl. Neurobiol. *9:* 175–179 (1983).

Dr. C.Q. Mountjoy, St. Andrews Hospital, Northampton, NN1 5D6 (UK)

Interdiscipl. Topics Geront., vol. 25, pp. 90–100 (Karger, Basel 1988)

Morphometric Methods and Their Potential Value for Gerontological Brain Research

W. Meier-Ruge

Gerontological Brain Research Unit, Neuropathology Division,
Department of Pathology, University of Basel, Switzerland

The Central Position of Morphometry in Research on Neuroanatomy and Neuropathology

Morphometry has become an accepted tool of morphological research in the past 20 years as a result of numerous basic studies on stereological methodology [Weibel and Elias, 1967; Weibel, 1979]. The original manual morphometric methods (integration processes using counting eyepieces or grating point count methods [Weibel and Elias, 1967; Haug, 1972]) were time-consuming and required great accuracy and concentration on the part of the investigator.

The introduction of optical-electronic image analysis [e.g. Kamin et al., 1970; Serra, 1975; Miller et al., 1979; Brugal, 1984] has markedly reduced observer errors and greatly speeded up the process. Morphological findings can be substantiated by stereological techniques based on reliable and reproducible measured data [Weibel, 1979]. Structural changes in a tissue can be quantified by morphometry and correlated with biochemical and physiological data [e.g. Hunziker et al., 1974; Meier-Ruge and Schulz, 1980, 1987; Neary et al., 1986a–c, Sumpter et al., 1986].

Present-Day Morphometric Methods

Many optical-electronic devices have been developed in recent years and they have become increasingly sophisticated with the introduction of new electronic systems. Furthermore, the cost of such equipment has steadily fallen. As typical examples of the devices available we may mention the Leitz Texture Analyzer, the Zeiss IBAS 2000, the Kontron image analyzer, the 'Quantimet' 720 by Cambridge Instruments, and the Leitz ASBA. The ongoing development of more specialized and more rapid processors will doubtless improve image quality and thus lead to greater accuracy, so that in future histological sections will not have to be cut and stained to today's high standards.

All the instruments are based on the *recognition of the grey values* of histological structures. Their range has now been extended to include the *processing of colour information.* A precondition for the successful use of all the measurement processes is that the structures which are to be stereologically counted should display optimal contrast with their surroundings.

Latterly the scope of the measurement process has been extended by introduction of *class size discrimination.* Thus modern analyzers are now capable of distinguishing neurons from neuroglia by size class distribution when the neuronal perikarya have been stained with cresyl violet. Similarly senile plaques visualized by means of an acid phosphatase reaction may be distinguished from neurons and neuroglia by means of a discrimination process which eliminates structures significantly smaller or larger than senile plaques.

By computer techniques it is now possible to eliminate *background structures* completely, thereby enhancing specific structural features – a question of some importance in the stereological analysis of *electron microscopic images.* In recent years this problem has gained much importance, especially in stereological investigation of mitochondria and synapses revealed by phosphotungstic acid staining [Bertoni et al., 1986].

Depending upon the problem to be solved, the histological sections may be evaluated automatically or, as is more usual, interactively. Modern electronic measuring systems have an extremely high sensitivity range and can distinguish between up to 256 grey values in a sample under investigation. Exclusion of structures according to size and shape may be achieved by image-modifying operations. *Image transformation,* e.g. erosion, dilatation or closure of gaps in the structure (often arising due to the fact that parts of the structure to be measured have low grey values) makes it possible to exclude artefacts and to optimize the image. These aids can be best exploited in interactive semi-automatic counting techniques. The *measurement fields are normally randomly selected* using a randomization programme to control movements of the motor-driven x/y cross matrix bar. However, depending upon the features measured it is also possible to select the measurement fields consecutively or along a meandering path.

For analysis of simple images and images with strong contrasts (e.g. measurement of particle sizes in pharmacy, etc.) fully automatic measuring cycles are available.

To avoid the need for complicated calculations to correct for shrinkage artefacts we prefer to work mainly with native material, the specimens being further processed on the cryostat for sectioning.

The Use of Stereological Techniques in Gerontological Brain Research and Future Trends

Morphometry has made great strides especially in brain research. Stereology furnishes reproducible numerical data on purely neuroanatomical features, e.g. the differences between the capillary networks of the cerebral cortex and the putamen [Meier-Ruge and Schulz, 1980].

These techniques have also made it possible for the first time to compare shrinkage phenomena such as occur normally in the aging brain with shrinkage phenomena resulting from pathological atrophic changes in the brain.

Stereological examination of the capillary network of the white matter and the cerebral cortex provided a means of studying shrinkage phenomena affecting various brain structures under normal and pathological conditions. The parameters measured are the 3-dimensional constants of the capillary network. Owing to the rigidity of the capillary basal membrane and the minor fluctuations of the blood pressure in the capillaries, the measured parameters are relatively constant. It is therefore possible to measure changes in corticocerebral volume without taking into account the size or weight of the brain.

Measurements have revealed that factors which might be expected to affect stereological capillary network parameters, such as the *time elapsing between death and removal of the brain and autolysis*, are actually of little importance and can for the most part be neglected when working with autoptic brain tissue [Hunziker and Schweizer, 1977; Schulz et al., 1980]. Significant changes in capillary volume fractions and capillary lengths are observed only in the first 30 min after death, no further changes occurring for another 20 h [Hunziker and Schweizer, 1977]. Morphometry has provided an answer to the question of whether the narrowing of the gyri and deepening of the sulci resulted from changes in cortical volume. Until fairly recently it was thought that the narrowing of the gyri was due to

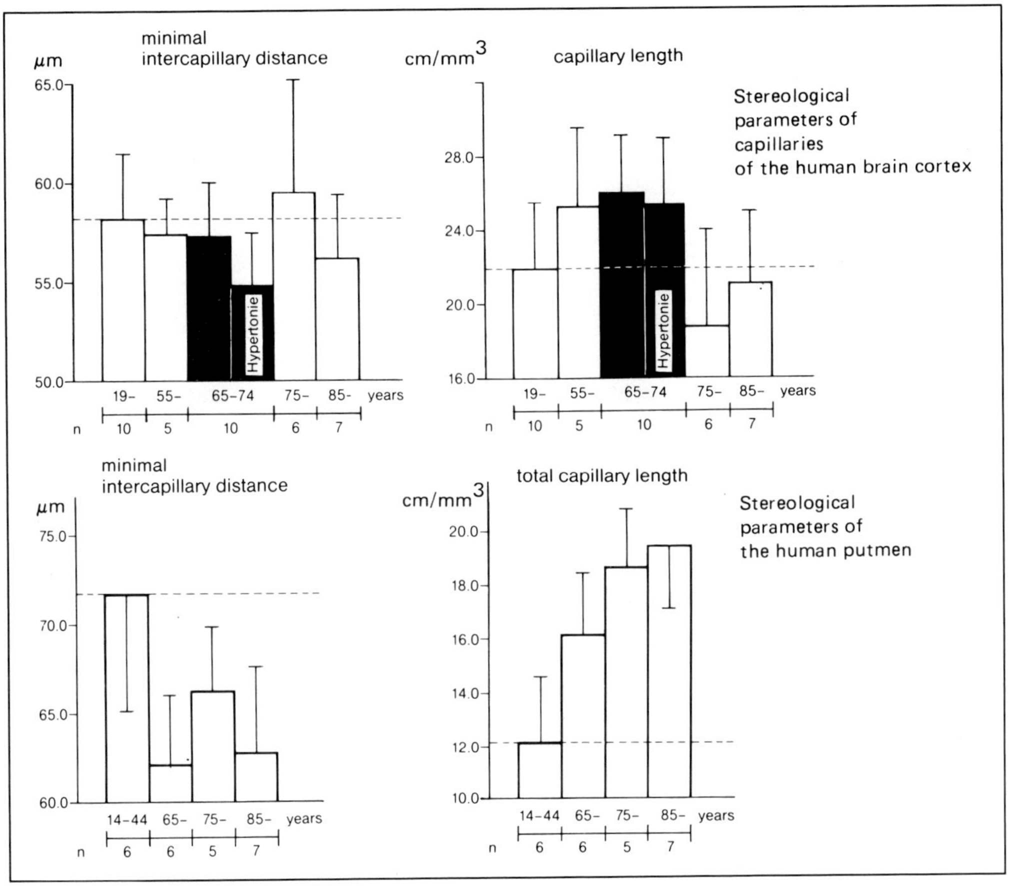

Fig. 1. Age-related change in capillary texture of the cerebral cortex and putamen. In normal aging the cerebral cortex does not display any marked shrinkage phenomena. On the other hand, the putamen, which is rich in white matter, undergoes a reduction in volume which manifests itself as an increase in capillary length and a decrease in intercapillary distance [Meier-Ruge et al., 1980].

age-related shrinkage of the cerebral cortex [Miller et al., 1980]. However, we were unable to show any age-related trend to increased corticocerebral capillary length and volume or decrease in corticocerebral intercapillary distance. In normal individuals the various capillary parameters of normal cerebral cortex remain within the normal range up to an advanced age, i.e. within the range of values for adults aged between 20 and 50 years (fig. 1).

This indicates that the cerebral cortex does not undergo any age-related shrinkage in the normal aging process [Meier-Ruge et al., 1980, 1984; Mann et al., 1986]. On the other hand, subcortical structures, such as the putamen, which are rich in white matter, display a significant age-dependent increase in the various capillary parameters [Meier-Ruge et al., 1984]. From the age of 75 onwards the increase in capillary length and decrease in intercapillary distance is less in evidence, although here also the volume shows an unmistakable tendency to decrease [Meier-Ruge et al., 1980, 1984]. Mann et al. [1986] arrived at quite similar findings.

Preliminary *investigations on subcortical white matter* underlying areas of cerebral cortex which are not subject to shrinkage phenomena have revealed capillary changes analogous to those seen in the putamen [Meier-Ruge, 1982]. Here again a marked reduction in volume was detectable from the age of 65 onwards (fig. 2).

Capillary diameters have been found to be relatively constant. The capillary diameter shows only minor changes, since owing to their rigid basal membrane the capillaries are capable of only limited dilatation or constriction. Indeed, capillary diameter can be used to gauge the accuracy of stereological measurements on capillaries. Morphometry also provides reliable information on changes in the size of nerve cells in the individual brain structures during normal and pathological aging. Since it is by no means easy to differentiate the perikarya of small neurons from glial cells, *absolute cell counts* must be viewed with some reservation. Depending upon the method employed, absolute cell counts tend to be overestimates or underestimates [Terry et al., 1981; Mountjoy et al., 1983; Hubbard and Anderson, 1985; Haug, 1972].

Nevertheless, in comparative studies it is quite possible, provided the methodological error remains constant, to obtain reliable data on changes in nerve cell count, nerve cell volume and nerve cell forms (determination of the form factor).

Measurement of neuronal volume provides important data on atrophic phenomena affecting the neurons, especially the corticocerebral neurons and avoids the problem of determining absolute neuron counts. Because we do not know exactly how large the cerebral cortex is, it is impossible to make a reliable estimate of the absolute number of cells it contains, and this is in any case ruled out by the different layers in its various regions. For this reason, measurement of neuronal volume has proved more reliable than cell counts as a means of assessing pathological changes in identical areas of the cerebral cortex.

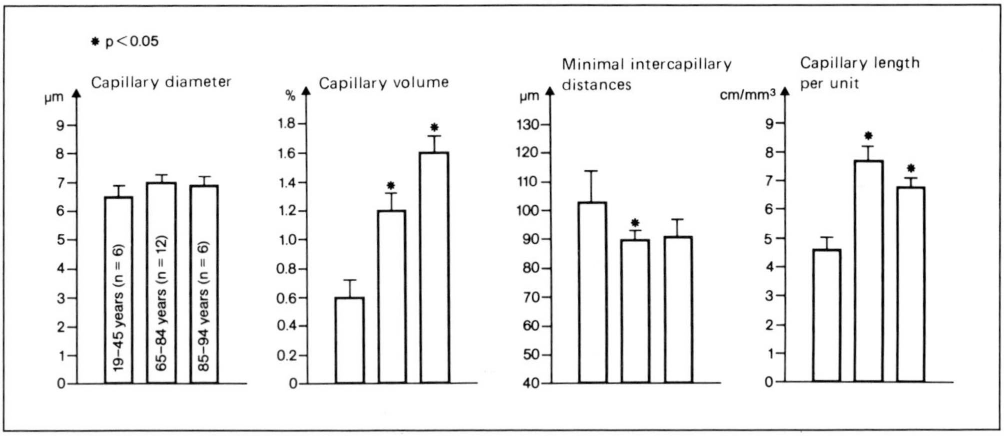

Fig. 2. Age-related change in subcortical white matter of the frontal, precentral and temporal gyrus (mean values of the capillary parameters). The decrease in volume of the white matter occurring after the age of 65 years results in a significant increase in capillary volume and capillary length and a reduction in intercapillary distance [Meier-Ruge, 1982].

The nerve cells of the cerebral cortex do not show any significant change in *size up to the age of 85 years.* A slight, but significant reduction in neuronal volume compared with the value for adults aged 19–45 years is seen under conditions of normal aging only in the 85- to 94-year age group. In normal aging, therefore, changes in neuronal volume are observed very late in life [Meier-Ruge et al., 1980; Haug, 1985]. Similarly in the putamen a significant reduction in neuronal volume is also seen very late in life, but is much more pronounced than the similar change in the cerebral cortex (fig. 3).

In brains form individuals with clinically and neuropathologically confirmed senile dementia, neuronal volume was 60% less than the value for controls of similar age (fig. 4) [Meier-Ruge et al., 1985]. Bondareff and Mountjoy [1986] have reported analogous findings for the locus coeruleus. Other authors have reported a loss of corticocerebral neurons [Terry et al., 1981; Mountjoy et al., 1983; Hubbard and Anderson, 1985].

The number and size of *senile plaques* may be measured morphometrically after selective staining with acid phosphatase. No correlation is discernible between neuronal atrophy and the number and size of senile plaques [Meier-Ruge et al., 1985]. Nor does plaque count or plaque size

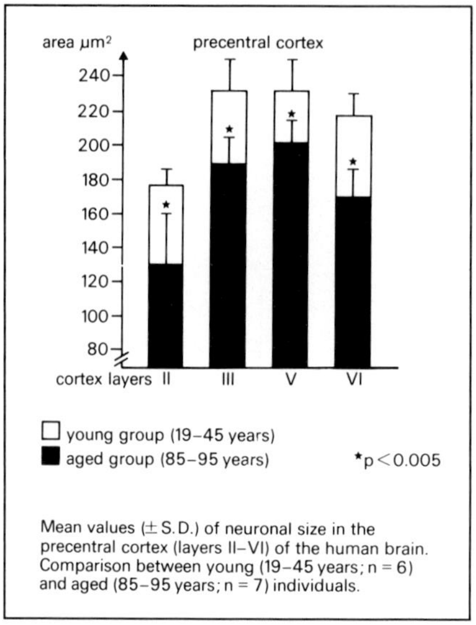

Fig. 3. Aging does not significantly affect neuronal volume in the various layers of the cerebral cortex until after the age of 85 years [Meier-Ruge et al., 1980].

correlate with reduction in neuronal volume. The change in volume of the cerebral cortex and putamen determined by stereological measurement of capillary parameters reveals that cerebrocortical atrophy is the predominant feature of senile dementia – no significant shrinkage phenomena of the type seen in the putamen can be demonstrated [Meier-Ruge et al., 1984].

Synaptic contact zones may be selectively visualized in electron micrographs by impregnation with phosphotungstic acid. *Synapses can then be submitted to morphometric analysis.* Such studies have revealed important findings on synaptogenesis and synaptic plasticity of the brain [Huttenlocher, 1979; Huttenlocher et al., 1982; Bertoni et al., 1986]. Cerebral aging is accompanied by a decrease in the number and an increase in the size of synaptic contact zones in the dentate gyrus [Cragg, 1975; Glick and Bondareff, 1979; Bertoni et al., 1985, 1986]. Other studies have contributed to a better understanding of synaptic changes occurring in cerebral disorders associated with dementia.

Stereology has also proved to be an important tool in experimental pathology for detection of pathophysiological changes [e.g. Hunziker et al.,

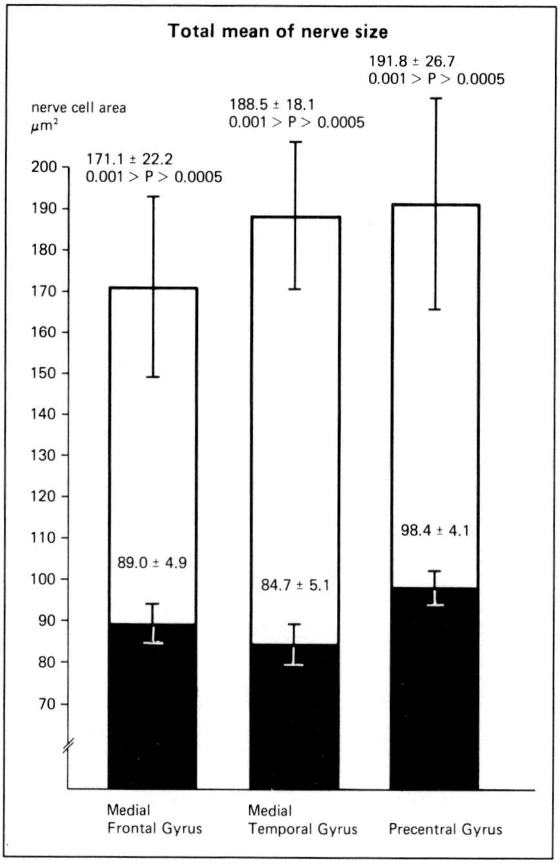

Fig. 4. Morphometric neuronal changes in senile dementia. Stereological investigations on demented brains reveal a 60% decrease in mean neuronal volume in the frontal, temporal and precentral gyrus [Meier-Ruge et al., 1985].

1974; Meier-Ruge and Schulz, 1987; Bertoni et al., 1985]. Stimulation of neuronal activity, which is accompanied by an increase in the microcirculation, is reflected in capillary stereology as an increase in capillary length and capillary volume, whereas depression of neuronal activity has the reverse effect [Meier-Ruge and Schulz, 1987]. In gerontological brain research, stereology furnishes much new important information on physiological and pathophysiological changes in the individual brain structures [Sumpter et al., 1986; Neary et al., 1986a–c].

Conclusion

With the introduction of stereological techniques in gerontological brain research, real progress has been made in the assessment of shrinkage processes affecting various brain structures and of changes in the volume of individual brain compartments.

Morphometry converts morphological findings to reproducible, numerical stereological values, such as number, length, perimeter, area and distance, permitting direct correlation with neurochemical and functional parameters.

Morphometry is an advance on descriptive morphology based solely on visual inspection of images, providing as it does an objective measure of spatial changes in structures within an organ. At the same time, morphometry also serves to measure morphological structural changes when no obvious pathological changes are visible on ordinary inspection, i.e. in specific conditions characterized only by gradual structural changes.

Under normal conditions the capillary density of the cerebral cortex is twice that of the putamen.

Age-related shrinkage of the cerebral cortex cannot be demonstrated in normal aging, but such shrinkage can be shown to occur in subcortical structures such as the putamen and subcortical brain tissue. In senile dementia, on the other hand, shrinkage of the cerebral cortex is a prominent feature.

Stereological investigation of normally aging brains has revealed a significant reduction in neuronal volume from the age of 85 years onwards. In senile dementia the volume of the corticocerebral neurons was reduced by 60%.

The rapid progress made in morphometry, especially the introduction of optical-electronic image analysis, has given fresh impetus to gerontological brain research.

References

Bertoni-Freddari, C.; Giuli, C.; Pieri, C.; Paci, D.: Quantitative investigation of the morphological plasticity of synaptic junctions in rat dentate gyrus during aging. Brain Res. *366:* 187–192 (1986).

Bertoni-Freddari, C.; Giuli, C.; Pieri, C.; Paci, D.; Dravid, A.: A modulating effect of Hydergine® on the synaptic plasticity of old rats. J. Pharmacol., Paris *16:* suppl. 3, pp. 33–38 (1985).

Bondareff, W.; Mountjoy, C.Q.: Numbers of neurons in nucleus locus ceruleus in demented and non-demented patients: Rapid estimation and correlated parameters. Neurobiol. Aging. *7:* 297–300 (1986).

Brugal, G.: Image analysis of microscopic preparations. Meth. Achiev. exp. Pathol., vol. 11, pp. 1–33 (Karger, Basel 1984).

Cragg, B.G.: The density of synapses and neurons in normal, mentally defective and ageing human brains. Brain *98:* 81–90 (1975).

Glick, R.; Bondareff, W.: Loss of synapses in the cerebellar cortex of the senescent rat. J. Geront. *34:* 818–822 (1979).

Haug, H.: Stereological methods in the analysis of neuronal parameters in the central nervous system. J. Microscopy *95:* 165–180 (1972).

Haug, H.: Are neurons of the human cerebral cortex really lost during aging? A morphometric examination; in Traber, Gispen, Senile dementia of the Alzheimer type, pp. 150–163 (Springer, Berlin 1985).

Hubbard, B.M.; Anderson, J.M.: Age-related variations in the neuron content of the cerebral cortex in senile dementia of Alzheimer type. Neuropath. appl. Neurobiol. *11:* 369–382 (1985).

Hunziker, O.; Frei, H.; Schulz, U.: Morphometric investigations of capillaries in the brain cortex of the cat. Brain Res. *65:* 1–11 (1974).

Hunziker, O.; Schweizer, A.: Post-mortem changes in stereological parameters of cerebral capillaries. Beitr. Pathol. *161:* 244–255 (1977).

Huttenlocher, P.R.: Synaptic density in human frontal cortex – developmental changes and effects of aging. Brain Res. *163:* 195–205 (1979).

Huttenlocher, P.R.; Courten, C. de; Garey, L.J.; Van der Loos, D.: Synaptogenesis in human visual cortex – evidence for synapse elimination during normal development. Neurosci. Lett. *33:* 247–252 (1982).

Kamin, G.; Kluge, N.; Müller, W.; Rzeznik, J.: Leitz-Classimat. Ein Instrument zur optischen Bilderfassung. Leitz-Mitt. Wiss. Technik, Suppl. Bd 1/7, pp. 1–24 (1970).

Mann, D.M.A.; Eaves, N.R.; Marcyniuk, B.; Yates, P.O.: Quantitative changes in cerebral cortical microvasculature in ageing and dementia. Neurobiol. Aging *7:* 321–330 (1986).

Meier-Ruge, W.: Age-induced changes of subcortical white matter – a morphometric investigation. Archs suisses Neurol. Neurochir. Psychiat. *131:* 249–250 (1982).

Meier-Ruge, W.; Hunziker, O.; Schulz, U.; Tobler, H.J.; Schweizer, A.: Stereologic changes in the capillary network and nerve cells of the aging human brain. Mech. Age. Dev. *14:* 233–243 (1980).

Meier-Ruge, W.; Schulz, U.: Kapilläre Veränderungen des Gehirns beim alternden Menschen und ihre pathophysiologische Bedeutung. Bull. schweiz. Akad. med. Wiss. *36:* 161–175 (1980).

Meier-Ruge, W.; Schulz-Dazzi, U.: Effects of Bromvincamine on the stereological parameters of corticocerebral capillaries. Life Sci. *40:* 943–949 (1987).

Meier-Ruge, W.; Ulrich, J.; Abdel-Al. S.: Stereologic findings in normal brain aging and Alzheimer's disease, in Wertheimer, Marois, Senile dementia: Outlook for the future, pp. 125–135 (Liss, New York 1984).

Meier-Ruge, W.; Ulrich, J.; Stähelin, H.B.: Morphometric investigation of nerve cells, neuropil and senile plaques in senile dementia of the Alzheimer type. Arch. Gerontol. Geriat. *4:* 219–229 (1985).

Miller, A.K.H.; Alston, R.L.; Corsellis, J.A.N.: The practical application of image-analysing systems to neuropathology: Principles and possible scope; in Smith, Cavanagh, Recent advances in neuropathology, pp. 113–127 (Churchill Livingstone, Edinburgh 1979).

Miller, A.K.H.; Alston, R.L.; Corsellis, J.A.N.: Variation with age in the volumes of grey and white matter in the cerebral hemispheres of man: measurements with an image analyser. Neuropath. appl. Neurol. *6:* 119–132 (1980).

Mountjoy, C.Q.; Roth, M.; Evans, N.J.R.; Evans, H.M.: Cortical neuronal counts in normal elderly controls and demented patients. Neurobiol. Ageing *4:* 1–11 (1983).

Neary, D.; Snowden, J.S.; Bowen, D.M.; Sims, N.R.; Mann, D.M.A.; Benton, J.S.; Northen, B.; Yates, P.O.; Davison, A.N.: Neuropsychological syndromes in presenile dementia due to cerebral atrophy. J. Neurol. Neurosurg. Psychiat. *49:* 163–174 (1986a).

Neary, D.; Snowden J.S.; Bowen, D.M.; Sims, N.R.; Mann, D.M.A.; Yates, P.O.; Davison, A.N.: Cerebral biopsy in the investigation of presenile dementia due to cerebral atrophy. J. Neurol. Neurosurg. Psychiat. *49:* 157–162 (1986b).

Neary, D.; Snowden, J.S.; Mann, D.M.W.; Bowen, D.M.; Sims, N.R.; Northen, B.; Yates, P.O.; Davison, A.N.: Alzheimer's disease: a correlative study. J. Neurol. Neurosurg. Psychiat. *49:* 129–137 (1986c).

Schulz, U.; Hunziker, O.; Frei, H.; Schweizer, A.: Postmortem changes in stereological parameters of cerebral neurons. Path. Res. Pract. *166:* 260–270 (1980).

Serra, J.: Mathematical morphology and image analysis (SIAM, Washington 1975).

Sumpter, P.Q.; Mann, D.M.A.; Davies, C.A.; Neary, D.; Snowden, J.S.; Yates, P.O.: A quantitative study of the ultrastructure of pyramidal neurons of the cerebral cortex in Alzheimer's disease in relationship to the degree of dementia. Neuropath. appl. Neurobiol. *12:* 321–329 (1986).

Terry, R.D.; Peck, A.; DeTeresa, R.; Schechter, M.D.; Horoupian, D.S.: Some morphometric aspects of the brain in senile dementia of the Alzheimer type. Ann. Neurol. *10:* 184–192 (1981).

Weibel, E.R.: Stereological methods (Academic Press, London 1979).

Weibel, E.R.; Elias, H.: Quantitative methods in morphology (Springer, Berlin 1967).

Prof. Dr. W. Meier-Ruge, Gerontological Brain Research Unit,
Neuropathology Division, Department of Pathology, University of Basel,
CH–4003 Basel (Switzerland)

III. The Neuronal Cytoskeleton in Senile Dementia of Alzheimer Type and Other Types of Dementia

Interdiscipl. Topics Geront., vol. 25, pp. 101–105 (Karger, Basel 1988)

Cytoskeletal Immunohistochemistry of Alzheimer Dementia and Related Diseases

J. Ulrich, A. Probst, B.H. Anderton, J.P. Brion

Institute of Pathology, University of Basel, Switzerland

A cytoskeleton can be observed in nearly all living cells. Ultrastructurally it consists of filamentous structures of three different aspects: microfilaments (diameter 6–7 nm), microtubules (outer diameter 25 nm) and intermediate filaments (diameter 8–11 nm). The microfilaments consist of actin (MW 41,800) and the microtubules of two subunits (MW 54,000 and 56,000, respectively) [Schliwa et al., 1986]. The intermediate filaments have a different chemical composition, depending on the cell type in which they occur [Anderton, 1981].

The intermediate filaments in the neurons are termed 'neurofilaments'. They are observed in the perikaryon, the axon and in the dendrites. Ultrastructurally they are characterized by side arms which connect the individual neurofilaments with each other or with the microtubules. Chemically, vertebrate neurofilaments consist of three different proteins: Light (L, MW approx. 68,000), medium (M, MW approx. 150,000) and heavy (H, MW approx. 210,000). The L component is localized in the core, the H and M components partly in the side arms and partly in the core. Neurofilaments in the axon are heavily phosphorylated, especially in their side arms, while those in the normal perikaryon and in the dendrites are not [Sternberger et al., 1983].

Periodic wispy projections on microtubules contain accessory proteins, microtubule-associated proteins (MAPs) or tau proteins, which induce polymerization and enhance stability of microtubules in vitro. Among MAPs we will have to discuss mainly MAP 2 which is localized in the growing dendrites [Matus et al., 1981] and MAP tau, which in normal neurons is localized in the axon [Binder et al., 1985].

In table I and in figures 1 and 2 we summarize the results of various studies, notably our own, made on the changes of Alzheimer's disease

Table I. Staining properties and immunohistochemical reactions of Alzheimer neurofibrillary tangles (NFT) and Pick bodies (PB)

Type of change	Reaction				
	amyloid stains	Gallyas silver iodide	PHF antisera	tau antisera	antibodies to phosphorylated neurofilament epitopes BF 10, RT 97, 8 D 8
NFT	+	+	+	+	+
PB	–	–	+	+	+

(fig. 1, 2) and Pick's disease (fig. 3, 4). Pick's disease [Pick, 1906] as Alzheimer's, is an organic dementia but its morphology is different: there is a circumscribed atrophy usually localized in the frontal and temporal lobe including the hippocampus. In the atrophied areas spherical neuronal inclusions (Pick bodies) in the cytoplasm are seen, which ultrastructurally consist of single straight filaments and not of paired helical filaments [cf. Probst et al., 1983; Rasool and Selkoe, 1985; Ulrich et al., 1987]. Furthermore, there are swollen chromatolytic neurons reminiscent of neurons whose axon was severed.

The most interesting finding shown in table I, also illustrated in fig. 1–4, is certainly that immunocytochemistry of the spherical inclusions in Pick's disease is nearly identical to that of Alzheimer tangles, in spite of the ultrastructural difference of straight filaments in Pick's disease and paired helical filaments in Alzheimer dementia. This identity is all the more puzzling as the two structures have different staining properties when stained for amyloid (NFT positive, Pick bodies negative) and in the silver iodide stain by Gallyas [1971]. The molecular basis of this identical immunocytochemistry in two ultrastructurally and tinctorically different structures is unknown.

The immunological identity of two morphologically different intraneuronal structures in two different dementing diseases makes one speculate that there might be a relationship between the clinical syndrome on the one hand and the particular immunocytochemistry on the other hand. We can attempt to explain the clinical similarity by the following hypotheses:

(1) The presence of Alzheimer tangles or of Pick bodies induces neuronal death. For the case of Alzheimer tangles such neuronal death has been documented [Probst et al., 1982; Ball et al., 1985]. As the cortical and hippocampal regions involved in Pick's disease become very atrophic, it is

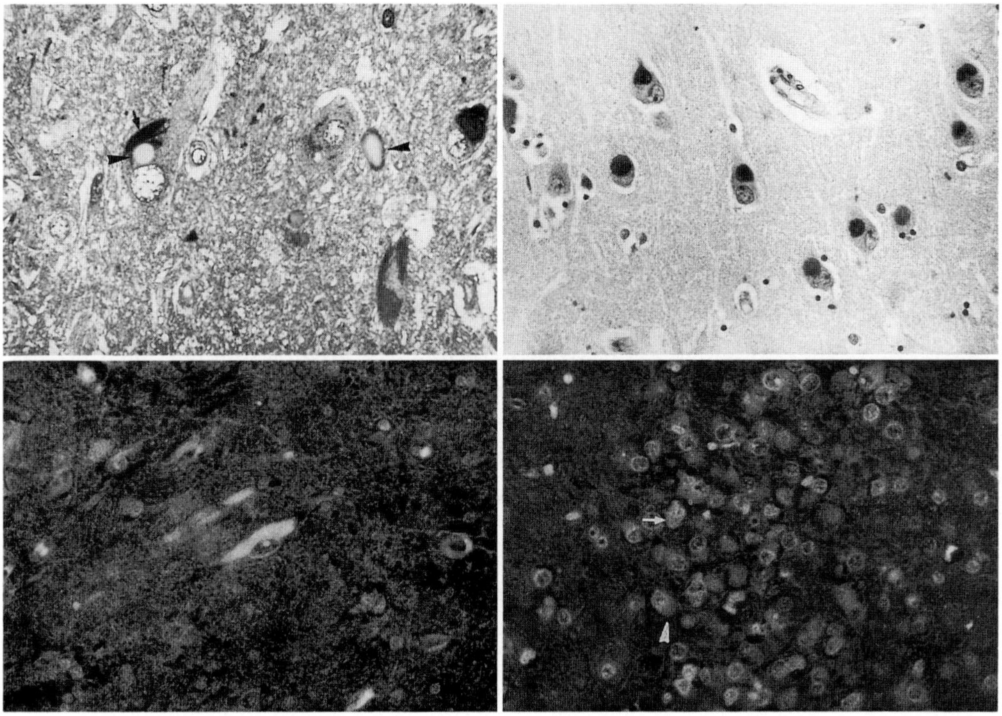

1 2 3 4

Fig. 1. A field in the hippocampus of a patient with Alzheimer dementia, immunostained with a monoclonal antibody to the neurofilament epitope M. One recognizes the dark Alzheimer neurofibrillary tangles (arrow) and white, unstained, Hirano bodies (wedges). 1-µmDurcupan-embedded sections. × 900.

Fig. 2. From the hippocampus of a patient with Pick's disease. Immunostained with the same monoclonal antibody as in figure. 1 (BF 10). One recognizes the dark Pick bodies in a similar staining intensity as Alzheimer neurofibrillary tangles in figure 1. Paraffin section. × 494.

Fig. 3. From a case with Alzheimer's disease (hippocampus). A neuron with Alzheimer neurofibrillary tangles is stained with thioflavine S for amyloid. Fluorescence microscopy on paraffin section. × 494.

Fig. 4. From the dentate gyrus of a patient with Pick's disease. Thioflavine S stain. Although nearly all neurons of this region contained Pick bodies (similar to figure 2), these remain invisible, however, with this stain. Possible localizations of Pick bodies at wedge and arrow. × 494.

very probable that here too the neurons die. Thus making this assumption, the syndrome of dementia could be explained simply by the neuronal loss, which occurs also in the hippocampus [cf. Ball, this volume]. In this case the syndrome of dementia would not be primarily related to common biochemical properties of the two types of neuronal inclusions.

(2) The cytoskeleton is responsible for the maintenance of the shape of a cell. In neurons, the shape is closely related to its function, electrophysiological properties being dependent on the geometry of axons and dendrites. A cytoskeletal alteration would probably induce an alteration of the cellular shape and thus change its functional properties. Making this second assumption one would relate dementia to a possible similarly altered neuronal geometry in both conditions, but not necessarily to the biochemical nature of Alzheimer tangles and Pick bodies.

(3) The chemical structure of the cytoskeleton is probably related to the surface specializations of the cell. So far, the evidence for this assumption is indirect: intermediate filaments are chemically different in epithelia, muscle cells, fibroblasts, neurons and glia. All these cell types are of course very different in their surface specializations, i.e. their receptors, antigens, intercellular contacts, etc. Furthermore, the cytoskeleton is chemically different in the various parts of the neuron: MAPs in the dendrites differ from those in the axons, phosphorylated epitopes are present in the normal axon but not in the perikaryon and in the dendrites. The processes of the neuron are obviously also very different from each other in their surface specializations: axons are often myelinated and have a smooth membrane, while dendrites tend to be covered with synaptic spines and other postsynaptic specializations. If there is a relationship between the chemistry of the cytoskeleton and the surface specialization, chemically similar neuronal inclusions as Pick bodies and Alzheimer tangles might be related to similar alterations of the neuronal surface, i.e. the synaptic organization, possibilities of transmitter release, etc. This similar alteration of the membranous structure in both diseases might be responsible for dementia.

It is a matter of future research to confirm or to refute one or the other of the three hypotheses. So far, neuronal death is proven in Alzheimer's disease and could probably be assessed in Pick's disease. The alteration in cell shape, especially of dendrites, was shown in Alzheimer's disease with Golgi techniques [Scheibel 1987; Probst et al., 1983] and is shown in the present volume by Chan-Palay by immunostaining of peptidergic neurons. Studies considering the third hypothesis, however, remain to be linked up with the studies of neurocytology on a molecular level.

References

Anderton, B.H.: Intermediate filaments. A family of homologous structures. J. Muscle Res. Cell Motil. *2:* 141–166 (1981).

Ball, M.J.; Hachinsky, V.; Fox, A.; Kissten, A.J.; Fisman, M.; Blume, W.; Kral, V.A.; Fox, H.; Merskley, H.: A new definition of Alzheimer's disease: a hippocampal dementia. Lancet *i:* 14–16 (1985).

Binder, L.I.; Frankfurter, A.; Rebhuhn, L.I.: The distribution of tau in the mammalian central nervous system. J. Cell Biol. *10:* 1371–1378 (1985).

Gallyas, F.: Silver staining of Alzheimer's neurofibrillary changes by means of physical development. Acta neuropath. hung. *19:* 1–8 (1971).

Matus, A.; Bernhardt, R.; Hugh-Jones, T.: High molecular weight microtubule associated proteins are preferentially associated with dendritic microtubules in the brain. Proc. natn. Acad. Sci. USA *78:* 3010–3014 (1981).

Pick, A.: Über einen weiteren Symptomen Komplex im Rahmen der Dementia senilis, bedingt durch umschriebene stärkere Hirnatrophie (gemischte Apraxie). Mschr. Psychiat. Neurol. *19:* 97–108 (1906).

Probst, A.; Basler, V.; Bron, B.; Ulrich, J.: Neuritic plaques in senile dementia of Alzheimer type. A Golgi analysis in the hippocampal region. Brain Res. *268:* 249–254 (1983b).

Probst, A.; Anderton, B.H.; Ulrich, J.; Kohler, R.; Kahn, J.; Heitz, P.U.: Pick's disease. An immunocytochemical study of neuronal changes. Acta neuropath. *60:* 175–182 (1983a).

Probst, A.; Ulrich, J.; Heitz, P.U.: Senile dementia of Alzheimer type. Astroglial reaction to extracellular neurofibrillary tangles in the hippocampus. An immunocytochemical and electron microscopic study. Acta neuropath. *57:* 75–79 (1982).

Rasool, C.G.; Selkoe, D.: Sharing of specific antigens by degenerating neurons in Pick's disease. An immunohistochemical study of neuronal changes. New Engl. J. Med. *312:* 70–705 (1985).

Scheibel, A.B.: Structural aspects of the aging brain: Spine system and the dendritic arbor; in Katzman, Terry, Bick, Alzheimer's disease: Senile dementia and related disorders, vol. 7: Aging, pp. 353–375 (Raven Press, New York, 1987).

Schliwa, M.: The cytoskeleton. An introductory survey. Cell biology monographs, vol. 1 (Springer, Wien 1986).

Sternberger, L.A.; Sternberger, N.H.: Monoclonal antibodies distinguish phosphorylated and nonphosphorylated forms of neurofilaments in situ. Proc. natn. Acad. Sci. USA *80:* 6126–6130 (1983).

Ulrich, J.; Haugh, M.; Anderton, B.H.; Probst, A.; Lautenschlager, C.; His, B.: Alzheimer dementia and Pick's disease: Neurofibrillary tangles and Pick bodies are associated with identical phosphorylated neurofilament epitopes. Acta neuropath. *73:* 240–246 (1987).

Prof. Dr. J. Ulrich, Institute of Pathology, University of Basel,
Schönbeinstrasse 40, CH–4056 Basel (Switzerland)

Interdiscipl. Topics Geront., vol. 25, pp. 106–118 (Karger, Basel 1988)

Structure and Chemistry of Alzheimer Neurofibrillary Tangles[1]

Brian H. Anderton[a], *Jean-Pierre Brion*[b], *Jaqueline Flament-Durand*[b], *Dominic Hue*[a], *Jacob Kahn*[c], *Christopher C.J. Miller*[a], *Alphonse Probst*[d], *Jürg Ulrich*[d]

[a] Department of Immunology, St. George's Hospital Medical School, University of London, UK; [b] Laboratoire d'Anatomie Pathologique et de Microscopie Electronique, Université Libre de Bruxelles, Belgium; [c] Department of Neuropathology, Institute of Psychiatry, University of London, UK; [d] Neuropathology Division, Institute of Pathology, University of Basel, Switzerland

Introduction

Neurofibrillary tangles, although not only found in Alzheimer's disease [Wisniewski et al., 1979], are present in large numbers in the brain of individuals who suffer from dementia of the Alzheimer type [Perry et al., 1978; Wilcock et al., 1982]. They are therefore one of the histopathological stigmata that are worthy of detailed analysis since elucidation of their structure and chemical composition will reveal something of the biochemical changes underlying their pathogenesis and this is likely to be linked to the process of neurodegeneration in Alzheimer's disease.

Neurofibrillary tangles are present in the perikarya of affected neurones. The distribution of the latter is described in detail in this volume in several articles, e.g. in that of Mountjoy et al. Tangles are large accumulations of fibrous material, the ultrastructure of which was first described by Kidd [1963, 1964] who showed them to contain predominantly bundles of filaments that are twisted about each other in pairs. These paired helical filaments (PHF) appear each to be comprised of two filaments approximately 10 nm wide wound around each other with a cross-over approxi-

[1] Supported by the Medical Research Council and the Wellcome Trust.

mately every 80 nm. Tangles also contain variable numbers of straight filaments, vesicular material and also amorphous material. PHF, however, are not exclusively found in perikaryal tangles but are also present intracellularly in the swollen abnormal neurites that contribute to the senile plaques [Gonatas et al., 1967; Kidd, 1963, 1964; Terry et al., 1964]. The anatomical relationship between PHF-containing neurites of the plaque and perikaryal tangles has yet to be determined. PHF have also recently been described in the telencephalic cortex of elderly and Alzheimer individuals; these neuropil threads of PHF were not associated with plaques or tangles [Braak et al., 1986].

PHF Structure

In figure 1 the ultrastructure of PHF is illustrated in situ (fig. 1a) and isolated (fig. 1b). Kidd [1963] first proposed that PHF may be abnormally arranged neurofilaments since the latter are 10 nm diameter filaments characteristic of neurones [Anderton, 1982; Lazarides, 1982]. Two more recent studies have come to the conclusion that the ultrastructure of PHF is inconsistent with their being simply normal neurofilaments twisted about each other in pairs [Crowther and Wischik, 1985; Wischik et al., 1985; Wisniewski and Wen, 1985]. Wisniewski and Wen [1985] have studied PHF ultrastructure by high resolution electron microscopy of PHF in ultrathin sections and have proposed a protofilamentous substructure that is different from that of neurofilaments. Crowther and Wischik [1985] and their colleagues [Wischik et al., 1985] have studied negatively stained preparations of isolated PHF and analysed these images by optical diffraction. They have proposed that PHF are not constructed from an extended fibrous protofilament as are neurofilaments but are built from a compact subunit stacked to form the two individual filaments of the pair in each PHF. There seems some disagreement in the handedness of the twist in PHF. Wischik et al. [1985] described the PHF to be wound in a left-handed helix, whereas Brion et al. [1984] described PHF with a right-handed twist. Wisniewski et al. [pers. commun.] apparently find that both left- and right-handed PHF can be found and the ratio varies from case to case of Alzheimer's disease. The significance of the handedness is not yet clear.

X-ray diffraction studies of PHF fractions have been reported to give a characteristic beta-pleated sheet structure which is consistent with their known amyloid-like staining properties with Congo red [Kirschner et al.,

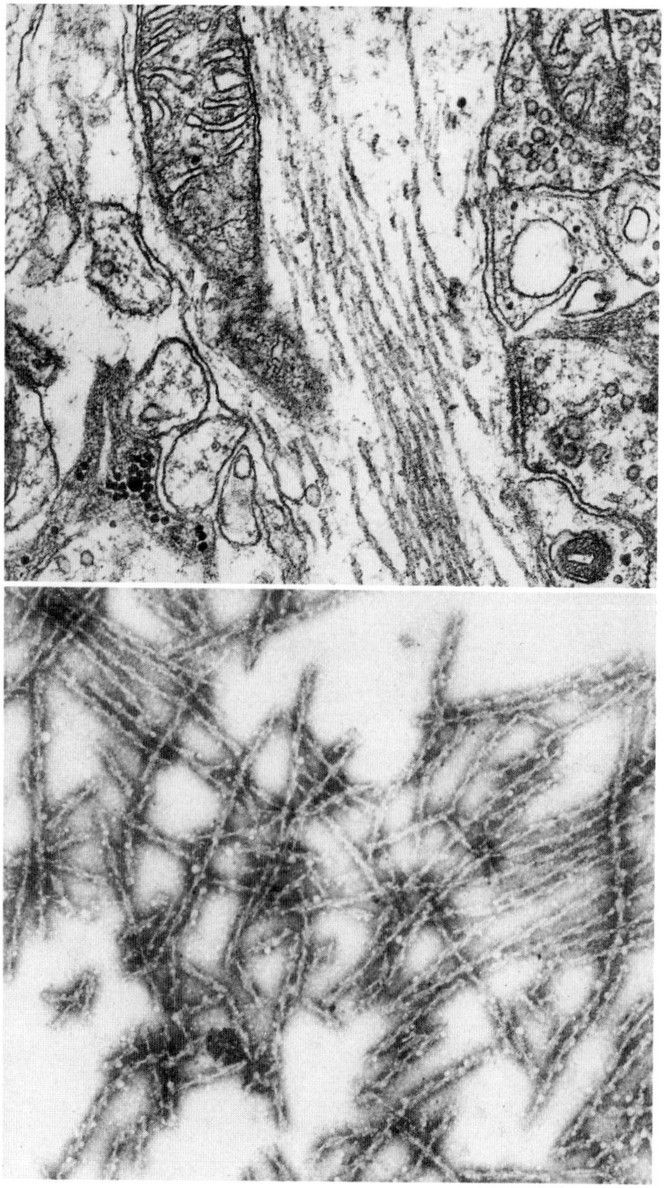

a

b

Fig. 1. *a* Electron micrograph of PHF in situ in a small neurite. × 70,000. *b* Electron micrograph of negatively stained isolated PHF, × 49,000.

1986]. Indeed, PHF are frequently classified as amyloid although this can lead to confusion with the extracellular amyloid deposits present in the senile plaques in Alzheimer's disease.

Chemical and Physical Properties of PHF

Selkoe et al. [1982] first reported that PHF appeared to be resistant to the usual protein solubilizing and denaturing agents and were therefore effectively insoluble. However, Iqbal et al. [1986a] have questioned whether or not PHF are insoluble and have claimed that extensive extraction of PHF with SDS and 2-mercaptoethanol does render them soluble. Others have recently reaffirmed the insolubility of PHF [Gorevic et al., 1986] and our own experience is that PHF resist solubilization [Hue and Anderton, unpubl. observations]. Iqbal et al. [1984] previously reported that there are differences in the ease of solubilizing PHF that are dependent upon the particular case of Alzheimer's disease and this should be investigated further since it may resolve the discrepancies between laboratories. Several polypeptides in the 47,000–70,000 molecular weight range were reported by the Iqbal group to be extracted from PHF and these may be derived from the microtubule-associated protein, tau [Grundke-Iqbal et al., 1986], that have been shown by immunocytochemical studies to be components of PHF (see below).

PHF, or at least a subpopulation of PHF, have been reported to be susceptible to proteolysis, particularly by proteinase K [Nukina and Ihara, 1985; Iqbal et al., 1986a]. Three prominent peptide fragments were obtained with molecular weights of 10,000, 26,000 and 36,000 [Nukina and Ihara, 1985].

Masters et al. [1985a, b] have claimed that a polypeptide of molecular weight 4,000, designated A4, is a common component of PHF and the extracellular amyloid. Amyloid in Alzheimer brain is found in the core of the senile plaques and in the walls of cerebral blood vessels. Glenner and Wong [1984] first isolated the vascular amyloid and obtained its partial sequence. The same sequence was obtained by Masters et al. [1985b] for the plaque core amyloid A4 protein. There is some doubt as to whether this A4 protein is also present in PHF since the purity of the PHF preparations used by Masters et al. [1985a] has been questioned [reviewed in Anderton and Miller, 1986]. In fact, Selkoe et al. [1986] were unable to repeat the work of Masters et al. [1985b] in that they were unable to obtain any sequence data on a polypeptide probably corresponding to A4 that they extracted from plaque cores isolated by flow cytometry; they suggested that their polypeptide has a blocked N-terminus. Selkoe et al. [1986] suggested that the PHF preparations of Masters et al. [1985a] were contaminated with vascular amyloid and that this amyloid is the source of A4 protein. Our own studies of PHF purification have led us to conclude that PHF are difficult to purify free of amyloid contamination, even using flow cytometry as the final stage of the PHF isolation protocol [Miller et al., 1986b]. Immunocytochemical studies have cast further doubt on the presence of A4 in PHF but have confirmed its presence in both plaque core and vascular amyloids (see below).

Immunocytochemical Studies of PHF

Neurofibrillary tangles in situ and their constituent PHF have been investigated with two types of antibody. Monoclonal and polyclonal antibodies have been produced to isolated fractions enriched in insoluble PHF; these reagents are usually referred to as anti-PHF antibodies. The other type of antibody that has been used to label PHF is one of known specificity and usually one raised to a known cytoskeletal protein, again these antibodies are both monoclonal and polyclonal. The cytoskeleton is the name given to the extensive fibrous network present in all eukaryotic cells and is comprised of three different fibrous organelles: 25 nm microtubules; 8–11 nm intermediate filaments which in neurones are represented by neurofilaments, and 6 nm microfilaments [Anderton, 1982]. Antibodies to many of the component proteins of these filamentous structures are now available.

Anti-PHF antibodies have been described by a number of different laboratories and all stain perikaryal tangles in situ and many abnormal neurites in the plaque periphery when studied by light microscopy [Brion et al., 1985a; Grundke-Iqbal et al., 1984; Ihara et al., 1983; Persuy et al., 1985; Wang et al., 1984; Yen et al., 1985]. These anti-PHF antibodies have generally been found not to stain any structures in healthy adult brain and this led to the supposition that PHF represent a unique structure, possibly the product of an infectious agent or a normally repressed gene. Anti-PHF antibodies have been shown by immunoelectron microscopy to label PHF in the tangle in situ and not other material in the tangle [Brion et al., 1985a; Miller et al., 1986a; Perry et al., 1986]. Neurofibrillary tangles in situ have been stained by some but not all antibodies to neurofilaments and to high molecular weight microtubule-associated proteins (HMWMAPs) [Anderton et al., 1982; Autilio-Gambetti et al., 1983; Elovaara et al., 1983; Gambetti et al., 1983; Haugh et al., 1986; Ihara et al., 1986; Kahn et al., 1980, 1987; Kosik et al., 1984; Miller et al., 1986a; Nukina and Ihara, 1986; Perry et al., 1986; Probst et al., 1983; Rasool et al., 1984; Rasool and Selkoe, 1984]. Several laboratories have now found that antibodies to another group of microtubule-associated proteins, designated tau, also stain tangles in situ; tau is a group of polypeptides in the 50,000–70,000 molecular weight range [Brion et al., 1985b; Delacourte and Defossez, 1986; Grundke-Iqbal et al., 1986a; Ihara et al., 1986; Kosik et al., 1986; Nukina and Ihara, 1986; Wood et al., 1986]. It is not clear from these reports if all antibodies to tau stain tangles in situ or if only a proportion of such antibodies are reactive with the tangles as is the case for neurofilament and HMWMAPs antibodies. Immunoelectron microscopic studies have shown that it is the PHF per se that are labelled by antibodies to neurofilaments and tau [Brion et al., 1985b; Miller et al., 1986a; Perry et al., 1986].

The labelling of isolated tangles by the antibodies described above is complex [Miller et al., 1986a]. We have studied several monoclonal antibodies to neurofilaments, a polyclonal antibody to PHF and a polyclonal antibody to tau. Of four neurofilament monoclonal antibodies that all label tangles in situ, one labels more than 90% of isolated tangles, two label a proportion of tangles that seem related to the case of Alzheimer brain used as the source of tangles and one labels virtually none of the isolated tangles. The polyclonal anti-PHF labels all isolated tangles, whereas the polyclonal anti-tau fails to label more than 10% of these isolated structures. Others [Kosik et al., 1986; Nukina and Ihara, 1986; Grundke-Iqbal et al., 1986b] have found that isolated tangles were labelled by anti-tau antibodies. These differences in findings suggest that either two populations of tau are associated with PHF with differing extents of post-translational modification or that the tau polypeptides may be proteolysed, some fragments readily lost, whilst others are stably incorporated into the insoluble PHF core. The results are summarized in table I and labelling by anti-PHF and one neurofilament monoclonal antibody of isolated PHF is illustrated in figure 2. These findings suggest that neurofilaments, at least, may be proteolysed when incorporated into PHF and that certainly isolated PHF are only partially representative of PHF in situ. We have shown that it is those fragments of neurofilament polypeptides that are responsible for the side arms

Table I. Pattern of labelling of neurofibrillary tangles by various antibodies: NF-H and NF-M are the two larger of the three neurofilament polypeptides with apparent molecular weights of 200,000 and 160,000, respectively

Antibody	Antigenic specificity	Tangles in situ	Isolated PHF, %
Monoclonals			
8D8	NF-H	+	>90
RT97	NF-H	+	20–50
1215	NF-H	+	20–50
BF10	NF-M	+	0
Polyclonals			
Anti-PHF	PHF and tau	+	>90
Anti-tau	tau	+	10

of neurofilaments that contribute to the PHF, and possibly the neurofilament core that has been degraded [Miller et al., 1986a].

We have found that our anti-PHF serum recognizes isolated tau proteins on Western blots in an identical pattern with an antiserum raised to purified tau proteins (fig. 2) [Brion et al., 1985c]. Others have also found that anti-PHF antisera contain antibodies to tau proteins [Delacourte and Defossez, 1986; Nukina and Ihara, 1986; Kosik et al., 1986; Grundke-Iqbal et al., 1986]. However, absorption of our anti-PHF antiserum with purified tau diminished but did not abolish staining of tangles nor did it prevent labelling of the insoluble excluded material at the top of the stacking gel on a Western blot of PHF although the labelling of tau polypeptides on Western blots was eliminated. Others have reported similar findings [Delacourte and Defossez, 1986; Nukina and Ihara, 1986; Kosik et al., 1986] although the extent of the reduction of tangle labelling by the tau-absorbed anti-PHF antisera was variable; Grundke-Iqbal et al. [1986] claimed that absorption with tau-containing microtubule proteins totally removed antibodies to PHF. It cannot be concluded what proportion of PHF is composed of tau since the above are not quantitative.

None of the various antibodies that label PHF have been shown to label amyloid and several antibodies to the isolated amyloid polypeptide have been demonstrated not to label PHF [Selkoe et al., 1986; Allsop et al., 1986]. Thus, at present there are no substantiated immunochemical data to suggest that amyloid and PHF are composed of the same protein.

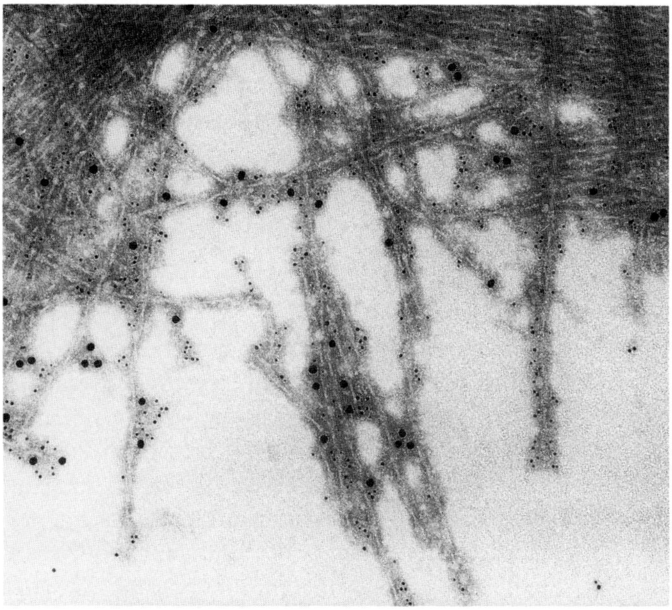

Fig. 2. Electron micrograph of isolated PHF double immunogold labelled with rabbit anti-PHF antibodies (large gold particles) and mouse monoclonal antibody, 8D8 (small gold particles). × 70,000.

PHF Are Phosphoproteins

We and others have analysed the specificity of some of the neurofilament antibodies that label tangles and it has been found that they are preferentially directed against phosphorylated neurofilament epitopes [Haugh et al., 1986; Sternberger et al., 1985; Cork et al., 1986]. This observation has led to the conclusion that there may be an abnormality in phosphoprotein metabolism in Alzheimer's disease because phosphorylated neurofilaments are found almost always only in axons whereas unphosphorylated neurofilaments are located in dendrites and perikarya [Sternberger and Sternberger, 1983].

A disturbance of phosphoprotein metabolism is also indicated by the observation that a monoclonal antibody to tau [Wood et al., 1986] only labels PHF after treatment with alkaline phosphatase [Grundke-Iqbal et al., 1986b]. Furthermore, tau proteins present in microtubule preparations isolated from Alzheimer brain are also only strongly recognized by this monoclonal antibody after dephosphorylation and microtubules assemble only poorly in extracts from Alzheimer brain and this may be the result of abnormally phosphorylated tau [Iqbal et al., 1986b].

Conclusion

The PHF of neurofibrillary tangles appear to be a composite of certain cytoskeletal proteins. Both tau and the neurofilament segment that are associated with tangles are located in the peripheral regions of microtubules and neurofilaments, respectively; neither is present in the filamentous cores of these two cytoskeletal organelles. Probably, tau and the neurofilament side arms are involved in linking microtubules and neurofilaments to adjacent cytoskeletal filaments or other organelles that are normally transported through the cytoplasm. Phosphorylation of neurofilaments and tau occurs normally and may have some role in regulating their, as yet unknown, functions. An abnormality of phosphorylation and possibly proteolysis may therefore be responsible for the disorganization of this cross-linking matrix of the cytoskeleton and consequent production of PHF. It is still unclear if PHF are only composed of these cytoskeletal proteins, albeit in an extremely modified form, or if other novel proteins are also present.

Study of the neurofilament and microtubule post-translational modifications are now important since the underlying mechanisms producing the neurofibrillary pathology are likely to involve these events.

References

Allsop, D.; Landon, M.; Kidd, M.; Lowe, J.S.; Reynolds, G.P.; Gardner, A.: Monoclonal antibodies raised against a subsequence of senile plaque core protein react with plaque cores, plaque periphery and cerebrovascular amyloid in Alzheimer's disease. Neurosci. Lett. *68:* 252–256 (1986).

Anderton, B.H.: The neuronal cytoskeleton: proteins and pathology; in Smith, Cavanagh, Recent advances in neuropathology, vol. 2, pp. 29–51 (Churchill Livingstone Edinburgh 1982).

Anderton, B.H.; Miller, C.: Proteins in a twist: Are neurofibrillary tangles and amyloid in

Alzheimer's disease composed of the same protein? Trends Neurosc. *9:* 337–338 (1986).

Autilio-Gambetti, L.; Gambetti, P.; Crane, R.C.: Paired helical filaments; relatedness to neurofilaments shown by silver staining and reactivity with monoclonal antibodies. Banbury Reports 15: Biochemical aspects of Alzheimer's disease, pp. 117–124 (Cold Spring Harbor Laboratory, Cold Spring Harbor 1983).

Braak, H.; Braak, E.; Grundke-Iqbal, I.; Iqbal, K.: Occurrence of neuropil threads in the senile human brain and in Alzheimer's disease: a third location of paired helical filaments outside of neurofibrillary tangles and neuritic plaques. Neurosci. Lett. *65:* 351–355 (1986).

Brion, J.-P.; Couck, A.M.; Flament-Durand, J.: Ultrastructural study of enriched fractions of 'tangles' from human patients with senile dementia of the Alzheimer type. Acta neuropath., Berl. *64:* 148–152 (1984).

Brion, J.-P.; Couck, A.M.; Passareiro, E.; Flament-Durand, J.: Neurofibrillary tangles of Alzheimer's disease: an immunocytochemical study. J. submicrosc. Cytol. *17:* 89–96 (1985a).

Brion, J.-P.; van den Bosch de Aguilar, P.; Flament-Durand, J.: Senile dementia of the Alzheimer type: morphological and immunocytochemical studies; in Traber, Gispen, Advances in applied neurological sciences; senile dementia of the Alzheimer type, pp. 164–174 (Springer, Berlin 1985b).

Brion, J.-P.; Passareiro, H.; Nunez, J.; Flament-Durand, J.: Mise en évidence immunologique de la protéine tau au niveau des lésions de dégénérescence neurofibrillaire de la maladie d'Alzheimer. Archs Biol., Brux. *96:* 229–235 (1985c).

Cork, L.C.; Sternberger, N.H.; Sternberger, L.A.; Casanova, M.F.; Struble, R.G.; Price, D.L.: Phosphorylated neurofilament antigens in neurofibrillary tangles in Alzheimer's disease. J. Neuropath. exp. Neurol. *45:* 56–64 (1986).

Crowther, R.A.; Wischik, C.M.: Image reconstruction of the Alzheimer paired helical filament. Eur. molec. Biol. Org. J. *4:* 3661–3665 (1985).

Delacourte, A.; Defossez, A.: Alzheimer's disease: tau proteins, the promoting factors of microtubule assembly are major components of paired helical filaments. J. neurol. Sci. *76:* 173–186 (1986).

Elovaara, I.; Paetau, A.; Lehto, V.-P.; Dahl, D.; Virtanen, I.; Palo, J.: Immunocytochemical studies of Alzheimer neuronal perikarya with intermediate filament antisera. J. neurol. Sci. *62:* 315–326 (1983).

Gambetti, P.; Shecket, G.; Ghetti, B.; Hiramo, A.; Dahl, D.: Neurofibrillary changes in human brain. An immunocytochemical study with a neurofilament antiserum. J. Neuropath. exp. Neurol. *42:* 69–79 (1983).

Glenner, G.G.; Wong, C.W.: Alzheimer's disease and Down's syndrome: sharing of a unique cerebrovascular amyloid fibril protein. Biochem. biophys. Res. Commun. *122:* 1131–1135 (1984).

Gonatas, N.K.; Anderson, W.; Evangelista, I.: The contribution of altered synapses in the senile plaque: an electron microscope study in Alzheimer's dementia. J. Neuropath. exp. Neurol. *26:* 25–39 (1967).

Gorevic, P.D.; Goni, F.; Pons-Estel, B.; Alvarez, F.; Peress, N.S.; Frangione, B.: Isolation and partial characterization of neurofibrillary tangles and amyloid plaque core in Alzheimer's disease: immunohistological studies. J. Neuropath. exp. Neurol. *45:* 647–664 (1986).

Grundke-Iqbal, I.; Iqbal, K.; Quinlan, M.; Tung, Y.-C.; Zaidi, M.S.; Wisniewski, H.M.: Microtubule-associated protein tau: a component of Alzheimer paired helical filaments. J. biol. Chem. *261:* 6084–6089 (1986a).

Grundke-Iqbal, I.; Iqbal, K.; Tung, Y.-C.; Quinlan, M.; Wisniewski, H.M.; Binder, L.I.: Abnormal phosphorylation of the microtubule-associated protein tau in Alzheimer cytoskeleton pathology. Proc. natn. Acad. Sci. USA *83:* 4913–4917 (1986b).

Haugh, M.C.; Probst, A.; Ulrich, J.; Kahn, J.; Anderton, B.H.: Alzheimer neurofibrillary tangles contain phosphorylated and hidden neurofilament epitopes. J. Neurol. Neurosurg. Psychiat. *49:* 1213–1220 (1986).

Ihara, Y.; Nukina, N.; Miura, R.; Ogawarr, M.: Phosphorylated tau proteins is integrated into paired helical filaments in Alzheimer's disease. J. Biochem. *99:* 1807–1810 (1986).

Ihara, Y.; Nukina, N.; Sugita, H.; Toyokura, Y.: Staining of Alzheimer's neurofibrillary tangles with antiserum against 200K components of neurofilament. Proc. Jap. Acad. *57B:* 152–156 (1981).

Iqbal, K.; Grundke-Iqbal, I.; Zaidi, T.; Ali, N.; Wisniewski, H.M.: Are Alzheimer neurofibrillary tangles insoluble polymers? Life Sci. *38:* 1695–1700 (1986a).

Iqbal, K.; Grundke-Iqbal, I.; Zaidi, T.; Merz, P.A.; Wen, G.Y.; Shaikh, S.S.; Wisniewski, H.M.; Alafuzoff, I.; Winblad, B.: Defective brain microtubule assembly in Alzheimer's disease. Lancet *ii:* 421–426 (1986b).

Iqbal, K.; Zaidi, T.; Thompson, C.H.; Merz, P.A.; Wisniewski, H.M.: Alzheimer paired helical filaments: bulk isolation, solubility and protein composition. Acta neuropath. *62:* 167–177 (1984).

Kahn, J.; Anderton, B.H.; Miller, C.C.J.; Wood, J.N.; Esiri, M.M.: Staining with monoclonal antibodies to neurofilaments distinguishes between subpopulations of neurofibrillary tangles, between groups of axons and between groups of dendrites. J. Neurol. (in press, 1987).

Kahn, J.; Green, P.G.; Thorpe, R.; Anderton, B.H.; Immunohistochemistry of neurofilaments in Alzheimer's disease. J. clin. exp. Geront. *2:* 199–210 (1980).

Kidd, M.: Paired helical filaments in electron microscopy of Alzheimer's disease. Nature, Lond. *197:* 192–193 (1963).

Kidd, M.; Alzheimer's disease – an electron microscopic study. Brain *87:* 307–321 (1964).

Kirschner, D.A.; Abraham, C.; Selkoe, D.J.: X-ray diffraction from intraneuronal paired helical filaments and extraneuronal amyloid fibrils in Alzheimer disease indicates cross beta-conformation. Proc. natn. Acad. Sci. USA *83:* 503–507 (1986).

Kosik, K.S.; Duffy, L.K.; Dowling, M.; Abraham, C.; McCluskey, A.; Selkoe, D.J.: Microtubule-associated protein. 2. Monoclonal antibodies demonstrate the selective incorporation of certain epitopes into Alzheimer neurofibrillary tangles. Proc. natn. Acad. Sci. USA *81:* 7941–7945 (1984).

Kosik, K.S.; Joachim, C.L.; Selkoe, D.J.: The microtubule-associated protein, tau, is a major antigenic component of paired helical filaments in Alzheimer's disease. Proc. natn. Acad. Sci. USA *83:* 4044–4048 (1986).

Masters, C.L.; Multhaup, G.; Simms, G.; Pottgiesser, J.; Martins, R.N.; Beyreuther, K.: Neuronal origin of a cerebral amyloid: neurofibrillary tangles of Alzheimer's disease contain the same protein as the amyloid plaque cores and blood vessels. Eur. molec. Biol. Org. J. *4:* 2757–2763 (1985a).

Masters, C.L.; Sims, G.; Weinmann, N.A.; Beyreuther, K.; Multhaup, G.; McDonald, B.L.: Amyloid plaque core protein in Alzheimer's disease and Down syndrome. Proc. natn. Acad. Sci. USA *82:* 4245–4249 (1985b).

Miller, C.J.; Brion, J.-P.; Calvert, R.; Chin, T.K.; Eagles, P.A.M.; Downes, M.J.; Flament-Durand, J.; Haugh, M.; Kahn, J.; Probst, A.; Ulrich, J.; Anderton, B.H.: Alzheimer's paired helical filaments share epitopes with neurofilament side arms. Eur. molec. Biol. Org. J. *5:* 269–276 (1986a).

Miller, C.J.; Haugh, M.; Anderton, B.H.; Hudson, L.; Murphy, J.: Neurofibrillary tangles from Alzheimer's disease brain purified using a cell sorter. Neurosci. Lett. *63:* 247–252 (1986b).

Nukina, N.; Ihara, Y.: One of the antigenic determinants of paired helical filaments is related to tau protein. J. Biochem. *99:* 1541–1544 (1986).

Perry, E.K.; Tomlinson, B.E.; Blessed, G.; Bergmann, K.; Gibson, P.H.; Perry, R.H.: Correlation of cholinergic abnormalities with senile plaques and mental test scores in senile dementia. Br. med. J. *ii:* 1457–1459 (1978).

Perry, G.; Selkoe, D.J.; Black, B.R.; Stewart, D.; Autilio-Gambetti, L.; Gambetti, P.: Electron microscopic localization of Alzheimer neurofibrillary tangle components recognised by an antiserum to paired helical filaments. J. Neuropath. exp. Neurol. *45:* 161–168 (1986).

Persuy, P.; Defossez, A.; Delacourte, A.; Tramu, G.; Bouchez, B.; Arnott, G.: Anti-PHF antibodies: an immunohistochemical marker of the lesions of Alzheimer's disease. Virchows Arch. *407:* 13–73 (1985).

Probst A.; Anderton, B.H.; Ulrich, J.; Kohler, R.; Kahn, J.; Heitz, P.U.: Pick's disease: an immunocytochemical study of neuronal changes. Monoclonal antibodies show that Pick bodies share antigenic determinants with neurofibrillary tangles and neurofilaments. Acta neuropath. *60:* 175–182 (1983).

Rasool, C.G.; Selkoe, D.J.: Alzheimer's disease: exposure of neurofilament immunoreactivity in SDS-insoluble paired helical filaments. Brain Res. *322:* 194–198 (1984).

Rasool, C.G.; Abraham, C.; Anderton, B.H.; Haugh, M.; Kahn, J.; Selkoe, D.J.: Alzheimer's disease: immunoreactivity of neurofibrillary tangles with anti-neurofilament and anti-paired helical filament antibodies. Brain Res. *310:* 249–260 (1984).

Selkoe, D.J.; Abraham, C.R.; Podlinsny, M.B.; Duffy, L.K.: Isolation of low molecular weight proteins from amyloid plaque fibres in Alzheimer's disease. J. Neurochem. *46:* 1820–1834 (1986).

Selkoe, D.J.; Ihara, Y.; Salazar, F.J.: Alzheimer's disease: insolubility of partially purified paired helical filaments in sodium dodecyl sulphate and urea. Science *215:* 1243–1245 (1982).

Sternberger, L.A.; Sternberger, N.H.: Monoclonal antibodies distinguish phosphorylated and non-phosphorylated forms of neurofilaments in situ. Proc. natn. Acad. Sci. USA *80:* 6126–6130 (1983).

Sternberger, N.H.; Sternberger, L.A.; Ulrich, J.: Aberrant neurofilament phosphorylation in Alzheimer's disease. Proc. natn. Acad. Sci. USA *82:* 4274–4276 (1985).

Terry, R.D.; Gonatas, N.F.; Weiss, M.: Ultrastructural studies in Alzheimer's presenile dementia. Am. J. Path. *44:* 269–297 (1964).

Wilcock, G.K.; Esiri, M.; Bowen, D.M.; Smith, C.T.: Alzheimer disease. Correlation of cortical choline acetyltransferase activity with severity of dementia and histological abnormalities. J. neurol. Sci. *57:* 407–417 (1982).

Wischik, C.M.; Crowther, R.A.; Stewart, M.; Roth, M.: Subunit structure of paired helical filaments in Alzheimer's disease. J. Cell Biol. *100:* 1905–1912 (1985).
Wisniewski, H.M.; Wen, G.Y.: Substructure of paired helical filaments from Alzheimer's disease neurofibrillary tangles. Acta neuropath., Berlin *66:* 173–176 (1985).
Wisniewski, K.; Jervis, G.A.; Moretz, R.C.; Wisniewski, H.M.: Alzheimer neurofibrillary tangles in diseases other than senile and presenile dementia. Ann. Neurol. *5:* 288–294 (1979).
Wood, J.G.; Mirra, S.S.; Pollock, N.J.; Binder, L.I.: Neurofibrillary tangles of Alzheimer's disease share antigenic determinants with the axonal microtubule-associated protein tau. Proc. natn. Acad. Sci. USA *83:* 4040–4043 (1986).

Brian H. Anderton, MD, Department of Immunology,
St. George's Hospital Medical School, University of London, Cranmer Terrace,
London SW17 ORE (UK)

IV. Aetiological and Pathogenetic Considerations about Alzheimer Dementia Based on Morphology

Interdiscipl. Topics Geront., vol. 25, pp. 119–139 (Karger, Basel 1988)

Typical and Atypical Viruses in the Aetiology of Senile Dementia of the Alzheimer Type

M.M. Esiri

Department of Neuropathology, The Radcliffe Infirmary, Oxford, UK

Introduction

The discoveries of classical virology established two important characteristics of typical virus diseases: firstly, the presence in the lesions of a filterable agent capable of transmitting the disease to another animal of the same or another species, and secondly, evidence of an inflammatory cell and immune response to the agent in the host. Furthermore, the diseases themselves were clearly infectious. Investigations of senile dementia of the Alzheimer type (SDAT) have so far failed convincingly to demonstrate any of these features and it might, therefore, be thought that the suggestion that viruses may be involved in its aetiology is implausible. However, investigations of central nervous system (CNS) diseases during the past 25 years have thrown up examples of indubitably virus-associated human diseases (and even more in animals) which lack one or other of these classical features, forcing us to contemplate previously unsuspected forms of virus-cell interactions in the generation of CNS disease (table I). A relatively simple example is progressive multifocal leucoencephalopathy (PML) caused by the JC polyoma virus, which produces a conventional lytic infection of oligodendrocytes and consequent demyelinating disease. In this case, the inflammatory and immune responses are greatly reduced or absent because there is immunosuppression in the host caused by concurrent disease, such as Hodgkin's disease. In another human disease, subacute sclerosing panencephalitis (SSPE), caused by measles virus, though there is evidence of inflammation and an immune response to measles, the transmissibility of the disease is difficult to demonstrate because the virus

Table I. Examples from human CNS viral diseases of variability in readiness with which transmissibility and evidence of inflammation/immunity can be shown

Typical viruses	Disease produced	Transmissibility	Evidence of inflammation/ immune response
Poliovirus	acute poliomyelitis	+++	+++
Arboviruses	acute encephalitis	+++	+++
Herpes simplex (whole virus)	acute encephalitis	+++	+++
Herpes simplex (latent virus)	none recognised except recrudescent skin eruption when reactivated	after in vitro cultivation	+/–
JC polyoma virus	progressive multifocal leucoencephalopathy	+ (transforms host cells)	+/–
Measles virus	subacute sclerosing panencephalitis	+/–	++
Atypical viruses	CJD/kuru	+ long incubation	–

assembled in the brain is apparently deficient and therefore poorly infective. Certain other diseases, previously classified as degenerative diseases, such as Creutzfeldt-Jakob disease (CJD) and kuru, have been shown to be transmissible, though the agents involved remain incompletely characterized and differ strikingly in physical, chemical and biological properties from conventional or 'typical' viruses. In these diseases, as in the similar disease of sheep, scrapie, there is no evidence of inflammation or of an immune response to the agent, despite an intact immune system, and the manner in which damage is inflicted in the CNS remains unclear. Most recently, the possible role of retroviruses in the development of human dementia has received increasing attention in the light of neuropathological findings in cases of dementia associated with acquired immune deficiency syndrome (AIDS). It is already apparent from the study of such diseases that there are hitherto unsuspected facets to CNS-'virus' interactions that may eventually be found to operate also in SDAT. In addition, there is the possibility that 'typical' viruses capable of establishing latent infections, such as the herpes viruses, may be able to influence cellular metabolism in the CNS in such a way as to facilitate the development of age-related diseases such as SDAT. In this chapter an attempt will be made to summarize the evidence for and against the involvement of such typical and atypical viruses in SDAT. Such evidence as there is at present is scanty and circumstantial. The justification for undertaking such a review is that it may clarify ways in which more direct approaches to the important question of a possible viral aetiology for SDAT might be taken.

Investigative Approaches to the Possible Viral Aetiology of SDAT

Attempts to Detect a Transmissible Agent

The presence of a typical or atypical virus can be demonstrated by transmission experiments involving inoculation of pathological tissue into animals, by attempts to culture such agents in vitro, by detecting virus-specific antigens using immunoenzyme or immunofluorescence techniques, by searching for viral particles with electron microscopy or, most recently, by attempts to detect non-host-derived nucleic acid by solution or in situ hybridization techniques. Each of these methods has its advantages and disadvantages (table II), but as different viruses may be most readily demonstrated using one rather than another, there is merit in utilizing them all in the investigation of SDAT. Results of such studies to date are reviewed below. It is important to remember that successful demonstration of the viral aetiology of a disease may be highly dependent on adoption of the appropriate technique, and availability of material from an appropriate, sometimes early, stage of the disease. Failure as yet to detect the presence of a virus, typical or atypical, in SDAT, cannot be regarded as firm evidence that a virus is not involved in its aetiology.

Attempts to Detect an Inflammatory or Immune Response

Histological examination of diseased tissues can enable the time course and extent of any inflammatory reaction to be determined. In classical viral encephalitis this is manifested by the presence of neuronophagia, perivascular cuffing, and parenchymal infiltration by lymphocytes, plasma cells and macrophages. The presence of these features, though highly characteristic, is not of course specific for active viral infection, but indicates rather the presence and localization of an immune response. With the use of monoclonal antibodies reactive with antigens specific for

Table II. Summary of approaches used to detect viruses in human diseases

Method	Notes	Examples of use
Animal transmission experiments	May provide positive results when other methods fail. If incubation period prolonged, animal may not live long enough to develop disease. May fail if animal host is not permissive for growth of agent. May fail if transmission attempted at stage of illness when phase of viral replication passed.	Most reliable method of detecting atypical viruses
In vitro cultivation	Many viruses are fastidious about type of cell used for cultivation. May fail if cultures attempted after phase in disease of viral replication. Can be used with explant culture and cell fusion to demonstrate latent infections. Provides good opportunity to study properties of the virus.	To detect many typical viruses, e.g. poliovirus, herpes simplex; explantation culture required for latent herpes simplex, measles in SSPE
Viral antigen detection in pathological tissues	Requires availability of high titre antibody to suspected virus. May require use of frozen sections. May fail if tissue taken from disease state after viral replication. Bypasses potential pitfalls in attempting to culture the virus or transmit it to animals.	Measles virus detected in SSPE: JC polyoma virus in PML: herpes simplex virus in encephalitis
Electron microscopy	Requires well-fixed material; relatively insensitive. Structural details that it provides are important clue to the nature of the virus.	Gave first clue to viral nature of SSPE and PML; can detect atypical virus-associated amyloid-like fibrils in brain extracts
Nucleic acid detection	May detect latent as well as active infection. Can be used on formalin-fixed material.	Has been used to detect herpes simplex genome in latent and reactivated state; HIV in AIDS encephalopathy; JC virus in PML

subclasses of inflammatory cells, for example, lymphocyte subsets, macrophages and the classes of immunoglobulin present in plasma cell cytoplasm, the inflammatory infiltrate can be characterized with some precision, but such an exercise unfortunately casts no light on the nature of the stimulating antigens. This is more readily accomplished by seeking evidence for specific humoral antibody production in serum and, particularly, cerebrospinal fluid (CSF), where it is reflected in an elevated immunoglobulin content and the presence of oligoclonal bands demonstrable on CSF electrophoresis. The presence of specific viral antibodies can be shown using such techniques as complement fixation, haemagglutination, neutralization and antibody-mediated cytotoxicity. Similarly, there are a variety of in vitro techniques available for analysing the reactivity of circulating white blood cells to viral antigens. However, these provide only a very indirect reflection of possible antiviral immune events that may be occurring in the CNS.

Other Approaches

The above techniques have proved themselves of value in definitive studies investigating the viral nature of many CNS diseases. There are other less direct approaches that are nevertheless also likely to be of value in investigating the possible viral nature of SDAT. These include epidemiological studies, pathological comparisons at both macroscopical and microscopical levels between SDAT and diseases of viral aetiology, and certain biochemical investigations such as detection of proteins, whether they be host or virus-derived, which are present in CNS or CSF only in the presence of typical or atypical virus infections. Recent examples of the detection of such proteins are to be found in the studies of lectin-binding glycoproteins in brain and CSF in CJD [Manuelidis et al., 1985], and of abnormal CSF proteins detectable by gel electrophoresis in CJD and herpes simplex encephalitis [Harrington et al., 1986]. In addition, the investigation of basic biochemical abnormalities, for example in cytoskeletal proteins, in SDAT on the one hand, and in virus-infected cells or tissues on the other, may uncover alterations that are common to these two situations.

Evidence for and against a Role for Typical Viruses in the Aetiology of SDAT

Two features of the pathology of SDAT support the view that viruses may be involved in its aetiology. The first concerns the distribution of the lesions, particularly the neurofibrillary tangles (NFT). These are found not randomly distributed in the brain, but concentrated in greatest profusion in areas of the limbic system [Hooper and Vogel, 1976; Ball, 1982], in neurons that lie in close connection with incoming fibres of the olfactory system. Primary olfactory (periamygdaloid) cortex, alone among the areas of primary sensory cortex, is severely affected by the pathological changes of SDAT [Pearson et al., 1985; Esiri et al., 1986], and in the amygdala the nuclei connected to olfactory structures bear the brunt of the pathology [Hertzog and Kemper, 1980]. The anterior olfactory nuclei also contain NFT [Esiri and Wilcock, 1984], and preliminary clinical testing has shown olfactory deficits in patients with SDAT [Corwin et al., 1985; Peabody and Tinklenberg, 1985; Serby et al., 1985]. This emphasis of the pathology in the olfactory pathways and areas connected closely with them raises the possibility that an environmental factor reaching the brain by the olfactory route may play a part in the development of SDAT. Viruses having ready access to the olfactory system from the nasopharynx or face are obvious candidates to consider in this context, though not the only ones.

The second pathological feature that suggests the possible involvement of viruses in the aetiology of SDAT is the NFT themselves. This curious form of intraneuronal pathological change is seen in some diseases of certain or presumed viral origin: herpes simplex encephalitis [Krücke, 1960], SSPE [Gutewa and Osetowska, 1961; Mandybur et al., 1977; Paula-Barbosa et al., 1979], and parkinsonism following encephalitis lethargica [Hallervorden, 1935]. However, it must be remembered that NFT have also been described in other conditions such as parkinsonism-dementia complex of Guam, chronic head trauma in boxers and lead poisoning [Wisniewski et al., 1970]. It is perhaps of interest that in those conditions in which NFT occur in association with viral encephalitis they are found not during the acute stage of the disease, when inflammation is most severe and virus most abundant, but at a later stage when inflammation is reduced or absent and degeneration and gliosis are more in evidence.

Finally, with regard to general evidence in favour of the involvement of typical viruses in SDAT, reference should be made to a recent electron microscopical study of an unusually well-fixed cerebral biopsy taken relatively early in the course of SDAT in a 62-year-old woman with a 2-year history of progressive dementia [Sloper et al., 1986]. Abnormal tubular structures were described in cortical pyramidal neuron cell bodies and proximal dendrites, where they were associated with granular endoplasmic reticulum (fig. 1). Some of the neurons containing these structures also contained paired helical filaments of NFT. The appearances suggested the growth of tubular structures invaginating the endoplasmic reticulum accompanied by loss of the ribosomes of the reticulum. While the nature of these abnormal tubular structures was not identified, attention was drawn to their similarity to elongated forms of certain viruses, particularly rhabdoviruses. This study reinforces the need to examine further biopsies by

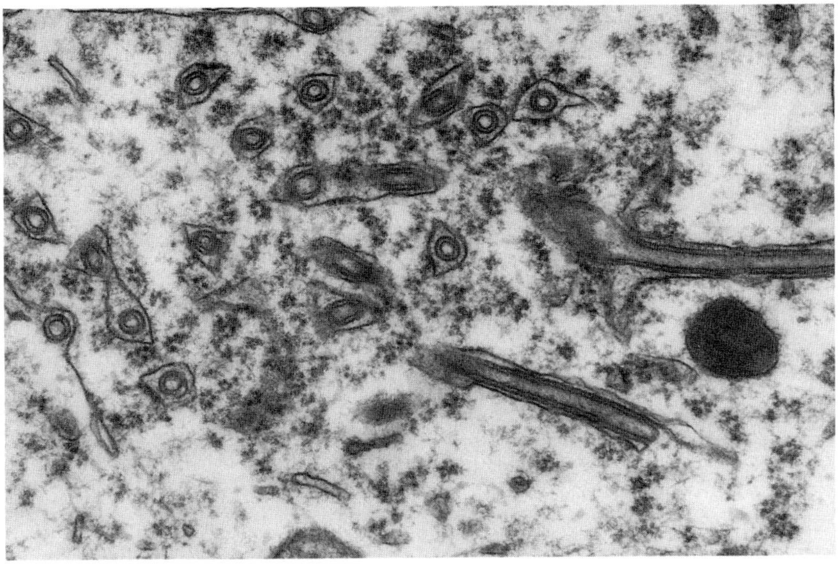

Fig. 1. Electron micrograph showing abnormal tubular profiles associated with the granular endoplasmic reticulum in a pyramidal neuron cell body in a cerebral cortical biopsy from a 62-year-old woman with Alzheimer's disease. × 37,400. (Courtesy Dr. J.J. Sloper.)

detailed electron microscopy and the other techniques outlined above whenever the opportunity arises.

Against the involvement of typical viruses in SDAT are: (1) the absence of neuronophagia and any marked inflammatory reaction in the brain; (2) the absence of elevated levels of immunoglobulin or oligoclonal immunoglobulin in CSF in amounts greater than age-matched controls [Jonker et al., 1982; Chu et al., 1983; Chapel et al., 1984]; (3) the absence of elevated antibody titres to common viruses in serum or CSF in SDAT [Whalley et al., 1980; Renvoize and Hambling, 1984], but see below for herpes simplex virus (HSV), and (4) the failure of attempts to transmit SDAT to experimental animals using brain material [Goudsmit et al., 1980; Corsellis, 1986]. These negative findings, while not ruling out the possibility of viral involvement in SDAT, suggest that, if typical viruses are involved, they are likely to be acting in a way that does not entail widespread, overt viral expression and multiplication. It is partly for this reason that the group of typical viruses that has come under closest scrutiny for a possible aetiological role in SDAT has been the herpes group, particularly

HSV, because of its known ability to establish latent infections in neurons. The evidence regarding this virus in relation to SDAT is reviewed below.

Herpes simplex Virus

HSV is a DNA virus of the herpes group which occurs in two slightly differing types, 1 and 2, and whose natural host is man. Infection with the type 1 virus is acquired by many people by the time they reach adult life. The type 2 virus causes a venereal infection and is not thought to be implicated in SDAT. Primary infection with type 1 virus may be unnoticed or may give rise to a stomatitis. Experimental studies in animals have established that, following primary infection, the virus enters nerve fibres at the portal of entry and travels along the axons until the sensory ganglion cells are reached, usually in the trigeminal ganglion. Here it may establish a latent infection with persistence of the genome which is detectable by nucleic acid in situ hybridization techniques [Galloway et al., 1979], but with little or no production of viral proteins, nor apparent alteration in neuronal metabolism. However, reactivation of virus occurs from time to time, and then whole virus is again synthesized, travels down the nerve and causes a focal eruption at the border of the lip or nasal mucosa. In humans, reactivation occurs under poorly understood conditions such as fever, emotional stress and exposure to strong sunlight. In both humans and experimental animals, section of the proximal nerve root attached to a latently infected ganglion regularly results in reactivation of the virus [Blyth and Hill, 1984].

In addition to producing a latent infection of the trigeminal (and certain other) ganglia, capable of intermittent reactivation, HSV occasionally produces a severe acute encephalitis which predominantly affects the limbic system and temporal lobes of the brain. In this form of encephalitis the virus lytically infects neurons and glial cells, virus can be readily cultured from the brain, and viral antigen can be demonstrated immunohistologically during the first 3 weeks of the disease. The areas of greatest damage in herpes simplex encephalitis closely parallel those containing the most severe lesions in SDAT [Hooper and Vogel, 1976; Ball, 1982]: hippocampus, parahippocampal gyrus, inferior and middle temporal cortex, insula, inferior parts of the frontal lobes, anterior cingulate cortex, amygdala and periamygdaloid cortex. However, the distribution of antigen within these regions in herpes simplex encephalitis does not precisely correspond with that of NFT in SDAT. For example, both dentate granule cells and pyramidal neurons in the hippocampus are heavily infected in herpes simplex encephalitis, but pyramidal cells, not dentate granule cells, contain NFT in SDAT; and in association cortex in herpes simplex encephalitis almost all the neurons seem susceptible to infection, whereas in SDAT, NFT in association cortex are chiefly located in pyramidal neurons of layers 3 and 5 [Pearson et al., 1985].

The distribution of damage in herpes simplex encephalitis, coupled with the well-known tendency of the virus to travel along neural pathways, and its capacity to cross synapses [Kristensson et al., 1982], leads to the suggestion that the virus may reach the brain to cause encephalitis by travelling along olfactory pathways [Johnson and Mims, 1968]. Viral antigen can indeed be demonstrated in human olfactory tracts and bulbs in herpes simplex encephalitis [Dinn, 1980; Ojeda, 1980; Esiri, 1982a]. Similarly, in SDAT the anterior olfactory nuclei, which in humans lie deep within the bulbs, contain NFT, thus extending the anatomical parallels between these two diseases from the medial temporal/limbic structures into more peripheral parts of the olfactory system. What is not at all clear in herpes simplex encephalitis is whether, if this view is correct, the virus travels along the olfactory route at the start of the illness, as suggested by Twomey et al. [1979], or whether it enters the brain this way asymptomatically at a much earlier time (perhaps at the time of primary infection), and establishes itself as a latent infection in parts of the brain related to central olfactory connections, from which it is reactivated by unknown factors to cause encephalitis. In mice it has been shown that experimental infection of the facial skin, cornea or nasal mucosa can result in HSV reaching the brain along both olfactory and trigeminal routes [Tomlinson and Esiri, 1983; Esiri and Tomlinson, 1984], and HSV-specific DNA has been detected in mice in olfactory/hippocampal neurons as well as in the brain stem after resolution of the acute infection following corneal inoculation [Stroop et al., 1984]. However, it has not been shown that HSV-specific DNA can be detected in olfactory regions of the brain in humans.

It should be added that some opinions favour other explanations than the olfactory route of entry of HSV for the distribution of the damage in herpes simplex encephalitis. Some consider it more likely that HSV reaches the brain by travelling along branches of the trigeminal nerve that innervate the leptomeninges covering the inferior frontal and temporal

Table III. Nucleic acid hybridization studies on detection of HSV in human brain

Technique	Material studied/disease/site	Results	Reference
In situ hybridization	brain tissue (site unspecified) from 3 cases of dementia and 3 of psychosis	+2 of 3 cases of dementia +1 of 3 cases of psychosis	Sequiera et al., 1979
Solution hybridization	brain tissue (selected specified sites) from 9 cases of Parkinson's disease and 1 control	all negative	Wetmur et al., 1979
Solution hybridization	brain tissue (plaque and non-plaque) from 5 cases of multiple sclerosis	all negative	Akulah et al., 1980
Solution hybridization	frontal and temporal lobe from 5 cases of dementia and 1 control	all negative	Middleton et al., 1980
Filter hybridization	brain tissue (some sites specified) from 5 cases of multiple sclerosis and 6 controls	+ in 4 of 5 cases of multiple sclerosis + in 3 of 6 controls	Fraser et al., 1981
Spot hybridization	brain tissue (multiple specified cortical and subcortical sites) from 8 cases of Alzheimer's disease and 9 controls	all negative	Taylor et al., 1984
In situ hybridization	brain tissue (site unspecified) from 17 cases of herpes encephalitis and 13 controls	+ in 16 of 17 cases of herpes encephalitis controls negative	Forghani et al., 1985
Dot blot hybridization	brain tissue (various specified sites) from 6 cases of epilepsy, 1 of parkinsonism, 1 of depression and 2 controls	+ in 5 of 6 cases of epilepsy + in 1 case of depression + in 1 case of parkinsonism + in 1 elderly control	Gannicliffe et al., 1985
Solution hybridization	brain tissue (selected but unspecified sites) from 18 cases of Alzheimer's disease and 5 controls	all negative	Pogo and Elizan, 1985
In situ hybridization	temporal lobe from unspecified number of cases of epilepsy, multiple sclerosis, Alzheimer's, Parkinson's and Huntington's diseases	+ in epilepsy remainder negative	Tourtellotte et al., 1985
In situ hybridization	temporal lobe smears from 7 immuno-compromised subjects, 1 case of herpes encephalitis and 3 controls	+ in 4 of 7 immuno-compromised subjects and the case of herpes encephalitis	Saldanha et al., 1986
In situ hybridization	brain tissue (site unspecified) from 25 cases of suspected herpes encephalitis	+ in 7 of 7 cases of proven herpes encephalitis remainder negative	Unger et al., 1986

lobes [Davis and Johnson, 1979; Ball, 1982], while others believe that the selectivity of the damage for the temporal lobes is more likely to be related to neurochemical peculiarities of the neurons in this region [Damasio and van Hoesen, 1985].

The whole important question of whether HSV establishes a latent infection in the brain in a large or small proportion of humans, whether such an infection is established at specific sites and whether it might be related to susceptibility to SDAT (and other psychiatric disease) is unresolved at present. The evidence, such as it is, is conflicting. It is derived principally from studies on the detection of HSV-specific nucleic acid. Such studies need to take account of the fact that some of the nucleic acid sequences present in the herpes genome are shared in common with mammalian cells [Puga et al., 1982; Peden et al., 1982]. Therefore, it is important that the labelled probes used to detect the HSV-specific DNA have been prepared to be complementary only to the sequences that are not so shared, and that adequate controls are employed. The studies that have examined human brain material for HSV-specific nucleic acid are listed in table III. Material has been examined in the various studies from normal brain and cases of SDAT, epilepsy, multiple sclerosis, Parkinson's disease, various psychoses and herpes simplex encephalitis. Solution and in situ hybridization techniques have variously been applied. The precise anatomical localization of the tissues examined has been specified in only a minority of studies, and in only one study was an olfactory structure (the olfactory tubercle) examined [Taylor et al., 1984]. Summarizing the results of the seven studies in which solution hybridization techniques were used, five gave entirely negative results and two gave positive results in some (but not all) cases of multiple sclerosis, epilepsy or controls. Of the five studies using in situ hybridization, two gave positive results in controls, while three produced positive results in some (but not all) cases of dementia, psychosis, epilepsy or compromised immunological states. Further studies will be required on anatomically defined brain regions including olfactory bulbs, hippocampus, and trigeminal central connections before the discrepancies between the findings in these studies can be resolved.

There seem to have been few attempts to detect latent HSV in human brain using the explantation and co-cultivation techniques that were originally used to demonstrate latency in sensory ganglia [Baringer and Swoveland, 1973; Stevens, 1975]. Nagington et al. [1976] reported isolation of HSV after co-cultivation of temporal lobe tissue for 18 days but that was from a case of suspected herpes simplex encephalitis. A report on the CNS infections encountered in subjects with AIDS quoted successful isolation of HSV from the brain in 4 cases. Two of these had subacute, progressive encephalomyelitis, 1 had chronic neurological dysfunction apparently due to a leucoencephalopathy and the last had no neurological disease [Levy et al., 1985]. The experience of Cabrera et al. [1980] suggests that explantation culture of CNS tissues is less sensitive than nucleic acid hybridization techniques for detecting HSV in brain and that explantation and co-cultivation of CNS tissue is a much less successful method of detecting latent HSV in brain than in ganglia. These workers found that explantation culture detected latent HSV in 95% of trigeminal ganglia 8 weeks after corneal inoculation of virus into mice. Brain explants yielded HSV from only 5% of explants whereas HSV-specific DNA was detected in 30% of brains.

Apart from the inconclusive evidence reviewed above on the detection of HSV-specific nucleic acid in the brain in SDAT, there are two suggestive, but far from conclusive reports of the occasional detection of HSV-specific antigen in human brains from subjects with SDAT [Mann et al., 1981, 1983; Esiri, 1982b]. Mann et al. [1981, 1983] studied temporal lobe cortex from 12 autopsy and 10 biopsy samples from cases of SDAT and 8 autopsy and 5 biopsy samples from controls using the PAP immunoperoxidase technique, and an HSV-specific antiserum. In one of the biopsy samples from a case of SDAT there was found a region of cerebral cortex containing HSV antigen in neurons and astrocytes. No associated inflammation was described. Using the same technique, Esiri [1982b] studied the hippocampus from 12 cases of SDAT as part of the control material examined during a study of viral antigen distribution in herpes simplex encephalitis. In 1 case of SDAT, HSV antigen was detected in the cytoplasm of three adjacent neurons in the dentate fascia granule cell layer (fig. 2). The cells harbouring the antigen did not appear necrotic and there was no associated inflammation. A subsequent study [unpublished] was undertaken of interrupted serial sections at 100-μm intervals through the full extent of both hippocampi from 6 cases of SDAT with entirely negative results. Roberts et al. [1986] also examined temporal lobe sections for HSV-specific antigen from 25 cases of SDAT and 32 controls, with negative results. It is difficult to know how to interpret the two positive findings. One possibility with respect to the autopsy material is that viral antigen expression represented reactivation of latent virus post mortem. Such an explanation would not, however, explain the finding of antigen in the biopsy specimen. It is perhaps conceivable that, if latent HSV is present in the human brain, some antigenic viral proteins (possibly, though not nec-

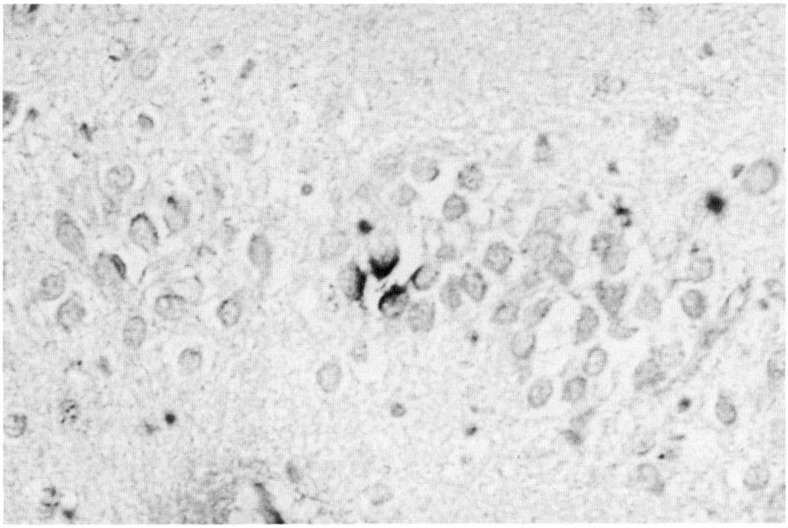

Fig. 2. Histological section from the dentate fascia of the hippocampus from a case of Alzheimer's disease treated with an antiserum to herpes simplex type I virus (Dako) using the immunoperoxidase technique. A focus of immunoreactivity is seen involving the cytoplasm of three adjacent neuron cell bodies. Counterstained with haematoxylin. ×400.

essarily, associated with complete infective virus production) may be focally produced from time to time on a scale that is sufficiently small to avoid evoking an immune response. Such low grade viral expression might be analogous to the slight viral shedding of HSV that can be detected in experimentally infected animals in the absence of overt recrudescent lesions [Blyth and Hill, 1984]. Whether such events, if they occur in the brain, might be associated with changes leading to SDAT, remains conjectural.

One additional study relating to the detection of possible HSV-related proteins in SDAT is the study of Wetmur et al. [1984] on DNA-binding proteins. These authors detected a DNA-binding protein in SDAT cerebral cortex which was not present in control samples and which had the same molecular weight (43,000) as a DNA-binding protein detectable in HSV-infected cells. Further characterization of these proteins to discover whether they have other common properties (apart from MW) would be of interest.

Evidence implicating HSV in SDAT on the basis of alterations in specific humoral immune responses is conflicting and will not be reviewed in detail here. Suffice it to say that raised antibody levels to HSV were found in CSF by Lycke [1984] and Libikova et al. [1975] in cases of SDAT, but not by Whalley et al. [1980].

There remain a few pointers which, while not amounting to firm evidence, support the view that some features of SDAT are not inconsistent with an aetiology related to HSV. One of these is the very common occurrence of HSV infection. Minor and localized expression of the pathological features of SDAT is so common in the aging brain that one is led to suspect that if a virus is involved in their evolution such a virus must be extremely widely distributed. Secondly, quite apart from the suggestive anatomical parallels between herpes simplex encephalitis and SDAT, the observation that the distribution of NFT in cortex supports the view that their development seems to extend along cortico-cortical connections [Pearson et al., 1985] would fit with the known spread of HSV along axons (though it is possible to envisage other factors doing likewise). Thirdly, as already stated, NFT have been described in herpes simplex encephalitis [Krücke, 1960], and fourthly, the suggestion from epidemiological studies [Heyman et al., 1984; Mortimer et al., 1985] that previous head injury, up to many years earlier than the onset of dementia, is more common in subjects with SDAT than age-matched controls, might be related to the known tendency for HSV to be reactivated by axonal trauma.

In summary, the evidence regarding a possible role for HSV in the aetiology of SDAT is inconclusive. Some investigators consider the evidence already available makes it unlikely that it is involved [Roberts et al., 1986]. My own view is that it is too early to dismiss the possibility and that further work is needed to explore the possibility of latent infection of the human brain by HSV, and to look for subtle alterations in cellular metabolism of latently infected neurons, particularly those that might influence the cytoskeleton.

Evidence for and against a Role for Atypical Viruses in the Aetiology of SDAT

Atypical virus infections of the CNS are characterized by being transmissible, after a prolonged incubation period, to experimental animals, and by the production of certain pathological changes in the brain. As already stated, SDAT has not yet been shown to be transmissible, despite initial claims that 2 cases of familial SDAT had been transmitted [Traub et al., 1977; Goudsmit et al., 1980; Corsellis et al., 1986]. This evidence must make it unlikely that SDAT is caused by an atypical virus infection unless its incubation period exceeds the life span of the animals in whom transmission experiments have been attempted. The mode of spread of the atypical virus agents to and within the CNS is thought to involve transport along axons [Kimberlin and Walker, 1980; Fraser and Dickinson, 1985], and in this respect spread of the agent would be consistent with the pattern of distribution of NFT in SDAT, though it is interesting that the limbic system is preferentially spared in CJD whereas it is preferentially damaged in SDAT [Karabetsos et al., 1985]. With regard to the pathological changes that occur in spongiform encephalopathies (such as CJD) associated with atypical virus infections, there are some differences and some similarities between them and SDAT. Widespread cortical neuron loss is usually conspicuous in the spongiform encephalopathies, but much less so in SDAT, except perhaps in the hippocampus. Spongy change, consisting of vacuoles 5–50 μm in diameter in grey matter, the most characteristic light microscopical change found in the spongiform encephalopathies, is not usually present in SDAT, although there may be a loosening of the cortical tissue in SDAT that may be difficult to distinguish from spongy change. However, the prominent vacuoles found in neuritic processes with electron microscopy in CJD have not been described in SDAT. Reactive gliosis is more prominent in the spongiform encephalopathies than in SDAT. NFT, which are so plentiful in SDAT, are not usually present in the spongiform encephalopathies. Despite these differences, which make the majority of cases of spongiform encephalopathy easily distinguishable on pathological grounds from the much more common cases of SDAT, there are occasional cases in which there is some overlap of pathology. Much interest has centered particularly on the fact that the argyrophilic, congophilic plaques, present in such abundance in SDAT, are also present, though not in identical form, in a minority of cases of spongiform encephalopathy, notably in kuru and a rare familial form of transmissible disease conforming clinically to the Gerstmann-Sträussler syndrome (dementia with spinocerebellar ataxia) [Masters et al., 1981] in humans, and in some forms of scrapie in mice [Fraser and Bruce, 1973]. The major part of these plaques consists of amyloid, and they show little of the corona of argyrophilic neuritic processes so commonly seen in SDAT (fig. 3). This common feature of amyloid deposition in the form of plaques in some cases of spongiform enceph-

a b c d
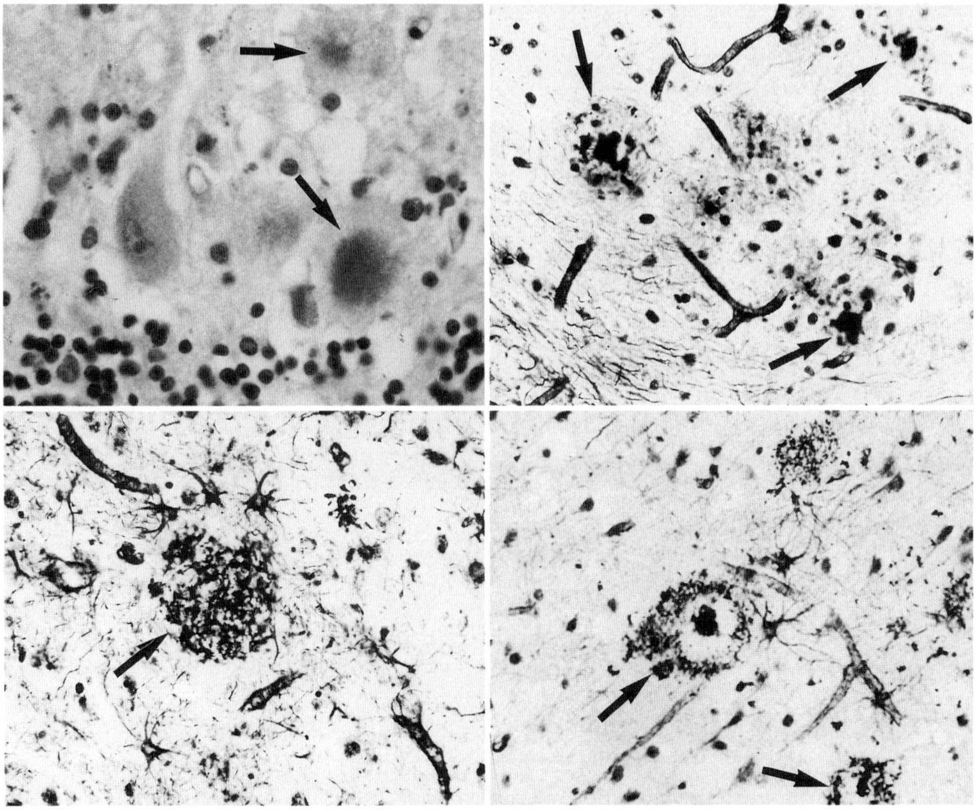

Fig. 3. Congophilic, argyrophilic plaques of spongiform encephalopathy and Alzheimer's disease compared. *a* Kuru (cerebellar cortex). *b* Gerstmann-Sträussler syndrome (cerebellar cortex). *c, d* Alzheimer's disease (cerebral cortex). Plaques indicated by arrows. *a* Congo red. × 374; *a–c* Von Braunmühl. × 190.

alopathy and all cases of SDAT has attracted additional interest since ultrastructurally identifiable amyloid-like fibrils have been regularly demonstrated in brain homogenates prepared from all types of spongiform encephalopathy, not just those that demonstrate amyloid plaques microscopically [Merz et al., 1983]. Such fibrils have not, however, been seen in material from cases of SDAT, plaque amyloid in this condition prepared in the same way appearing slightly different [Merz et al., 1982]. Moreover, amyloid deposition in the form of plaques and vascular infiltration, as occurs in SDAT, is found also in familial amyloid angiopathy, a disease which has not been transmitted and which is thought to be due to abnormal metabolism of a cysteine proteinase inhibitor, cystatin C, in the CNS [Grubb et al., 1984]. Thus, amyloid deposition in the brain seems more likely on present evidence to be a reactive change, and it would seem unsound to consider it a specific marker for the presence of atypical viruses [Carp et al., 1985; Somerville et al., 1985]. There is, therefore, very little evidence to implicate atypical viruses in SDAT at the present time.

Conclusion

The conclusion of this review is that the evidence for the involvement of viruses in the aetiology of SDAT is preliminary and inconclusive. On the other hand, it is certainly not justifiable to dismiss viruses as possible aetiological factors, and there are sufficiently suggestive findings in their favour to warrant further investigation. At the same time, other alternative aetiologies need to be explored, and these are reviewed in the succeeding chapters of this book.

References

Akulah, G.S.; Albrecht, P.; Tourtellotte, W.W.: Search for cytomegalovirus and herpes simplex virus genetic information in multiple sclerosis. Neurology *30:* 530–532 (1980).

Ball, M.: Limbic predilection in Alzheimers dementia: Is reactivated herpes virus involved? J. Can. Sci. Neurol. *9:* 303–306 (1982).

Baringer, J.; Swoveland, P.: Recovery of herpes-simplex virus from human trigeminal ganglions. New Engl. J. Med. *288:* 648–650 (1973).

Blyth, W.A.; Hill, T.J.: Establishment, maintenance and control of herpes simplex virus (HSV) latency; in Rouse, Lopez, The immunobiology of herpes simplex virus infection, pp. 10–32 (CRC Press, Boca Raton 1984).

Cabrera, C.; Wohlenberg, C.; Openshaw, H.; Rey-Mendez, M.; Puga, A.; Notkins, A.: Herpes simplex virus DNA sequences in the CNS of latently infected mice. Nature, Lond. *288:* 288–290 (1980).

Carp, R.; Merz, P.; Kascsak, R.; Merz, G.; Wisniewski, H.: Nature of the scrapie agent: current status of fact and hypotheses. J. gen. Virol. *66:* 1357–1368 (1985).

Chapel, H.; Esiri, M.M.; Wilcock, G.K.: Immunoglobulin and other proteins in the cerebrospinal fluid of patients with Alzheimer's disease. J. clin. Path. *37:* 697–699 (1984).

Chu, A.B.; Sever, J.L.; Madden, D.L.; Iivanainen, M.; Leon, M.; Wallen, W.; Brooks, B.R.; Lei, Y.J.; Houff, S.: Oligoclonal IgG bands in cerebrospinal fluid in various neurological diseases. Ann. Neurol. *13:* 434–439 (1983).

Corsellis, J.: The transmissibility of dementia. Br. med. Bull. *42:* 111–114 (1986).

Corwin, J.; Serby, M.; Conrad, P.; Rotrosen, J.: Olfactory recognition deficit in Alzheimer's and parkinsonian dementias. IRCS med. Sci. *13:* 260 (1985).

Damasio, A.R.; Hoesen, G.W. van: The limbic system and the localisation of herpes simplex encephalitis. J. Neurol. Neurosurg. Psychiat. *48:* 297–301 (1985).

Davis, L.; Johnson, R.: An explanation for the localisation of herpes simplex encephalitis? Ann. Neurol. *5:* 2–5 (1979).

Dinn, J.J.: Transolfactory spread of virus in herpes simplex encephalitis. Br. med. J. *ii:* 1392 (1980).

Esiri, M.M.: Herpes simplex encephalitis: an immunohistological study of the distribution of viral antigen within the brain. J. neurol. Sci. *54:* 209–226 (1982a).

Esiri, M.M.: Viruses and Alzheimer's disease. J. Neurol. Neurosurg. Psychiat. *45:* 759 (1982b).

Esiri, M.M.; Pearson, R.C.A.; Powell, T.P.S.: The cortex of the primary auditory area in Alzheimer's disease. Brain Res. *366:* 385–387 (1986).

Esiri, M.M.; Tomlinson, A.H.: Herpes simplex encephalitis: immunohistological demonstration of spread of virus via olfactory and trigeminal pathways after infection of facial skin in mice. J. neurol. Sci. *64:* 213–217 (1984).

Esiri, M.M.; Wilcock, G.K.: The olfactory bulbs in Alzheimer's disease. J. Neurol. Neurosurg. Psychiat. *47:* 56–60 (1984).

Forghani, B.; Dupuis, K.W.; Schmidt, N.J.: Rapid detection of herpes simplex virus DNA in human brain tissue by in situ hybridisation. J. clin. Microbiol. *22:* 656–658 (1985).

Fraser, H.; Bruce, M.E.: Argyrophilic plaques in mice inoculated with scrapie from particular sources. Lancet *i:* 617–618 (1973).

Fraser, H.; Dickinson, A.G.: Targeting of scrapie lesions and spread of agent via the retino-tectal projection. Brain Res. *346:* 32–41 (1985).

Fraser, N.W.; Lawrence, W.C.; Wroblewska, Z.; Gilden, D.H.; Koprowski, H.: Herpes simplex type I DNA in human brain tissue. Proc. natn. Acad. Sci. USA *78:* 6461–6465 (1981).

Galloway, D.A.; Fenoglio, C.; Schevchuk, M.; McDougall, J.K.: Detection of herpes simplex RNA in human sensory ganglia. Virology *95:* 265–268 (1979).

Gannicliffe, A.; Saldanha, J.A.; Itzhaki, R.F.; Sutton, R.N.P.: Herpes simplex viral DNA in temporal lobe epilepsy. Lancet *i:* 214–215 (1985).

Goudsmit, J.; Morrow, C.; Asher, D.: Evidence for and against the transmissibility of Alzheimer disease. Neurology *30:* 945–950 (1980).

Grubb, A.; Jensson, O.; Gudmundsson, G.; Arnason, A.; Lötberg, H.; Malm, J.: Abnormal metabolism of a γ-trace alkaline microprotein. The basic defect in hereditary cerebral haemorrhage with amyloidosis. New Engl. J. Med. *311:* 1547–1549 (1984).

Gutewa, J.; Osetowska, E.: A chronic form of subacute sclerosing panencephalitis (a case with a history of 5 years): clinical and pathological study; in van Bogaert, Radermecker, Hozay, Lowenthal, Encephalitides, pp. 386–404 (Elsevier, Amsterdam 1961).

Hallervorden, J.: Anatomische Untersuchungen zur Pathogenese des post-encephalitischen Parkinsonismus. Dt. Z. NervHeilk. *136:* 68–77 (1935).

Harrington, M.G.; Merril, C.R.; Asher, D.M.; Gajdusek, D.C.: Abnormal proteins in the cerebrospinal fluid of patients with Creutzfeldt-Jakob disease. New Engl. J. Med. *315:* 279–283 (1986).

Herzog, A.G.; Kemper, T.L.: Amygdaloid changes in aging and dementia. Archs Neurol. *37:* 625–629 (1980).

Heyman, A.; Wilkinson, W.E.; Stafford, J.A.; Helms, M.J.; Sigmon, A.H.; Weinberg, T.: Alzheimer's disease: a study of epidemiological aspects. Ann. Neurol. *15:* 335–341 (1984).

Hooper, M.; Vogel, F.: The limbic system in Alzheimer's disease. Am. J. Path. *85:* 1–19 (1976).

Johnson, R.T.; Mims, C.A.: Pathogenesis of viral infections of the nervous system. New Engl. J. Med. *278:* 23–30, 84–92 (1968).

Jonker, C.; Eikelenboom, P.; Tavenier, P.: Immunological indices in the cerebrospinal fluid of patients with presenile dementia of the Alzheimer type. Br. J. Psychiat. *140:* 44–49 (1982).

Karabetsos, A.; Doshi, B.; Dunkley, B.: The temporal lobe in Creutzfeldt-Jakob disease. Neuropath. appl. Neurobiol. *11:* 324 (1985).

Kimberlin, R.H.; Walker, C.A.: Pathogenesis of mouse scrapie: evidence for neural spread of infection to the CNS. J. gen. Virol. *51:* 183–187 (1980).

Kristensson, K.; Nennesmo, I.; Persson, L.; Lycke, E.: Neuron-to-neuron transmission of herpes simplex virus – transport of virus to brain stem nuclei. J. neurol. Sci. *54:* 149–156 (1982).

Krücke, W.: Herpes simplex Virus und Nervensystem. Jb. Max-Planck-Ges. Förd. Wiss., pp. 70–84 (1960).

Levy, R.M.; Bredesen, D.E.; Rosenblum, M.L.: Neurological manifestations of the acquired immunodeficiency syndrome (AIDS): experience at UCSF and review of the literature. J. Neurosurg. *62:* 475–495 (1985).

Libikova, H.; Pogady, J.; Wiedermann, V.: Search for herpetic antibodies in CSF in senile dementia and mental retardation. Acta virol., Praha *19:* 493–495 (1975).

Mandybur, T.; Nagpaul, A.; Pappas, Z.; Niklowitz, W.: Alzheimer neurofibrillary change in subacute sclerosing panencephalitis. Ann. Neurol. *1:* 103–107 (1977).

Mann, D.; Tinkler, A.; Yates, P.: Neurological disease and herpes simplex virus. Acta neuropath., Berl. *60:* 24–28 (1983).

Mann, D.; Yates, P.; Davies, J.; Hawkes, J.: Viruses, parkinsonism and Alzheimer's disease. J. Neurol. Neurosurg. Psychiat. *44:* 651 (1981).

Manuelidis, L.; Valley, S.; Manuelidis, E.E.: Specific proteins associated with Creutzfeldt-Jakob disease and scrapie share antigenic and carbohydrate determinants. Proc. natn. Acad. Sci. USA *82:* 4263–4267 (1985).

Masters, C.L.; Gajdusek, D.C.; Gibbs, C.J., Jr.: Creutzfeldt-Jakob disease virus isolations from Gerstmann-Sträussler syndrome with an analysis of the various forms of amyloid plaque deposition in the virus-induced spongiform encephalopathies. Brain *104:* 559–588 (1981).

McLennan, J.; Darby, G.: Herpes simplex latency: the cellular location of virus in dorsal root ganglia and the fate of the infected cell following virus activation. J. gen. Virol. *51:* 233–243 (1980).

Merz, P.; Somerville, R.; Wisniewski, H.: Abnormal fibrils in scrapie and senile dementia of Alzheimer type; in Court, Virus non conventionnels et affections du système nerveux central, pp. 259–281 (Masson, Paris 1982).

Merz, P.; Somerville, R.; Wisniewski, H.; Manuelidis, L.; Manuelidis, E.: Scrapie-associated fibrils in Creutzfeldt-Jakob disease. Nature, Lond. *306:* 474–476 (1983).

Middleton, P.; Petric, N.; Kozak, M.; Rewcastle, N.; Crapper-McLachlan, D.: Herpes simplex viral genome and senile and presenile dementias of Alzheimer and Pick. Lancet *i:* 1038 (1980).

Mortimer, J.A.; French, L.R.; Hutton, J.T.; Schuman, L.M.: Head injury as a risk factor for Alzheimer's disease. Neurology *35:* 264–267 (1985).

Nagington, J.; Rubinstein, D.; Anderson, J.: Herpes simplex encephalitis. Lancet *i:* 493 (1976).

Ojeda, V.J.: Fatal herpes simplex encephalitis with demonstration of virus in the olfactory pathway. Pathology *12:* 429–437 (1980).

Paula-Barbosa, M.M.; Brito, R.; Silva, C.A.; Faria, R.; Cruz, C.: Neurofibrillary changes in the cerebral cortex of a patient with subacute sclerosing panencephalitis. Acta neuropath. *48:* 157–160 (1979).

Peabody, C.; Tinklenberg, J.: Olfactory deficits and primary degenerative dementia. Am. J. Psychiat. *142:* 524–525 (1985).

Pearson, R.C.A.; Esiri, M.M.; Hiorns, R.W.; Wilcock, G.K.; Powell, T.P.S.: Anatomical correlates of the distribution of the pathological changes in the neocortex in Alzheimer disease. Proc. natn. Acad. Sci. USA *82:* 4531–4534 (1985).

Peden, K.; Mounts, P.; Hayward, G.S.: Homology between mammalian cell DNA sequences and human herpes virus genomes detected by a hybridisation procedure with high-complexity probe. Cell *31:* 71–80 (1982).

Pogo, B.G.T.; Elizan, T.S.: Search for viral DNA sequences in Alzheimer brain tissue. Lancet *i:* 978 (1985).

Puga, A.; Cantin, E.M.; Notkins, A.L.: Homology between murine and human cellular DNA sequences and the terminal repetition of the S component of herpes simplex virus type I DNA. Cell *31:* 81–87 (1982).

Renvoize, E.B.; Hambling, M.H.: Cytomegalovirus infection and Alzheimer's disease. Age Ageing *13:* 205–209 (1984).

Roberts, G.; Taylor, G.; Carter, G.; Johnson, J.; Bloxham, C.; Brown, R.; Crow, T.: Herpes simplex virus: a role in the aetiology of Alzheimer's disease? J. Neurol. Neurosurg. Psychiat. *49:* 216 (1986).

Saldanha, J.; Sutton, R.N.P.; Gannicliffe, A.; Farragher, B.; Itzhaki, R.F.: Detection of HSVI DNA by in situ hybridisation in human brain after immunosuppression. J. Neurol. Neurosurg. Psychiat. *49:* 613–619 (1986).

Sequiera, L.; Carrasco, L.; Curry, A.; Jennings, L.; Lord, M.; Sutton, R.: Detection of herpes simplex viral genome in brain tissue. Lancet *ii:* 609–612 (1979).

Serby, M.; Corwin, J.; Conrad, P.; Rotrosen, J.: Olfactory dysfunction in Alzheimer's disease and Parkinson's disease. Am. J. Psychiat. *142:* 781–782 (1985).

Sloper, J.J.; Barnard, R.O.; Eglin, R.P.; Powell, T.P.S.: Abnormal tubular structures associated with the granular endoplasmic reticulum of the neocortical neurons in a biopsy from a patient with Alzheimer's disease. Neuropath. appl. Neurobiol. *12:* 491–501 (1986).

Somerville, R.: Ultrastructural links between scrapie and Alzheimer's disease. Lancet *i:* 504–506 (1985).

Stevens, J.: Latent herpes simplex virus and the nervous system. Curr. Top. Microbiol. *70:* 31–50 (1975).

Stroop, W.G.; Rock, D.; Fraser, N.: Localisation of herpes simplex virus in the trigeminal and olfactory systems of the mouse central nervous system during acute and latent infection by in situ hybridisation. Lab. Invest. *51:* 27–38 (1984).

Taylor, G.; Crow, T.J.; Markakis, D.A.; Lofthouse, R.; Neeley, S.; Carter, G.I.: Herpes simplex and Alzheimer's disease: a search for virus DNA by spot hybridisation. J. Neurol. Neurosurg. Psychiat. *47:* 1061–1065 (1984).

Tomlinson, A.H.; Esiri, M.M.: Herpes simplex encephalitis: immunohistological demonstration of spread of virus via olfactory pathways in mice. J. neurol. Sci. *60:* 473–484 (1983).

Tourtellotte, W.W.; Shapshak, P.; Delgado-Escuera, A.; et al.: Problems in the detection of measles, herpes simplex-I, SV40 and JC virus nucleotide sequences using cloned probe in situ hybridisation in brains of patients with chronic neurological disorders: including Huntington's, Parkinson's and Alzheimer's diseases, multiple sclerosis and temporal lobe epilepsy. J. Neurol., suppl. 232, p. 159 (1985).

Traub, R.; Gajdusek, D.; Gibbs, C., Jr.: Transmissible virus dementia: the relation of transmissible spongiform encephalopathy to Creutzfeldt-Jakob disease; in Aging and dementia, pp. 91–172 (Spectrum, New York 1977).

Twomey, J.A.; Barker, C.M.; Robinson, G.; Howell, D.A.: Olfactory mucosa in herpes simplex encephalitis. J. Neurol. Neurosurg. Psychiat. *42:* 983–987 (1979).

Unger, E.R.; Budgeon, L.R.; Myerson, D.; Brigati, D.J.: Viral diagnosis by in situ hybridisation. Description of a rapid simplified colorimetric method. Am. J. surg. Path. *10:* 1–8 (1986).

Wetmur, J.G.; Casals, J.; Elizan, T.S.: DNA binding profiles in Alzheimer's disease. J. neurol. Sci. *66:* 201–208 (1984).

Wetmur, J.G.; Schwartz, J.; Elizan, T.S.: Nucleic acid homology studies of viral nucleic acids in idiopathic Parkinson's disease. Archs Neurol. *36:* 462–464 (1979).

Whalley, C.; Urbaniak, S.; Darg, C.; Peutherer, J.F.; Christie, J.E.: Histocompatibility antigens and antibodies to viral and other antigens in Alzheimer's presenile dementia. Acta psychiat. scand. *61:* 1–7 (1980).

Wisniewski, H.; Terry, R.D.; Hirano, A.: Neurofibrillary pathology. J. Neuropath. exp. Neurol. *29:* 163–176 (1970).

M.M. Esiri, MD, Department of Neuropathology, The Radcliffe Infirmary, Oxford, OX2 6NE (UK)

Interdiscipl. Topics Geront., vol. 25, pp. 140–155 (Karger, Basel 1988)

Association of Aluminium and Silicon with Neuropathological Changes in the Ageing Brain

J. Candy[a], *A. Oakley*[a], *D. Gauvreau*[b], *P. Chalker*[c], *H. Bishop*[c], *D. Moon*[c], *G. Staines*[d], *J. Edwardson*[a,1]

[a]MRC Neuroendocrinology Unit, Newcastle General Hospital, Newcastle upon Tyne, UK; [b]Institut National de la Recherche Scientifique, Montreal, Canada; [c]Materials Characterisation Service, Harwell Laboratory, Oxfordshire, UK, and [d]EM Service Unit, Department of Metallurgy, University of Newcastle, Newcastle upon Tyne, UK

Introduction

Although it is over 100 years since the neurotoxicity of aluminium was first described, it is with the relatively recent advent of sensitive analytical techniques (especially atomic absorption spectrometry and neutron activation analysis), that its potential toxicity, in man, has been recognized. Industrial exposure to high levels has, in rare cases, been linked with gross neurological changes [McLaughlin et al. 1962]. There is convincing evidence for the involvement of aluminium in the progressive, fatal encephalopathy in patients on long-term haemodialysis for chronic renal failure [Alfrey et al., 1976]. In susceptible animal species, aluminium produces disruption of short-term memory, focal neurological deficits and seizures [Crapper, 1976]. Aluminium also induces neurofibrillary tangles in several neuronal types and these resemble, at least at the light microscopic level, the neurofibrillary tangles in Alzheimer's disease [Klatzo et al., 1965]. At the electron microscopic level the aluminium-induced tangles consist of bundles of single 10-nm filaments [Terry and Pena, 1965], unlike the paired helical filaments found in neurofibrillary tangles in Alzheimer's disease [Kidd, 1963]. The cytoskeletal abnormalities produced by alumin-

[1] The authors are grateful to Mrs. D. Hinds for expert secretarial assistance and Ms. M. Moses, Mrs. J. Thompson and Mrs. M. Johnson for invaluable assistance.

ium stimulated investigations on the possible involvement of this metal ion in Alzheimer's disease. Initial reports that brain levels of aluminium were increased in Alzheimer's disease [Crapper et al., 1973, 1976] have since been disputed [Trapp et al., 1978; McDermott et al., 1979; Markesbery et al., 1981; Traub et al., 1981]. However, a recent report has claimed that markedly increased levels of aluminium were present in both the cerebral cortex and hippocampus in mainly presenile cases of Alzheimer's disease, when compared with age-matched controls [Ward and Mason, 1986]. These studies, on gross tissue samples, may be misleading since they do not allow focal tissue accumulations to be correlated with the discrete neuropathological changes that occur. Advanced microanalytical techniques (electron and proton microprobe X-ray microanalysis and secondary ion mass spectrometry), are now available that permit such direct correlative studies to be carried out. This chapter focuses attention on the distribution of aluminium and silicon revealed using these techniques in relation to neurofibrillary tangles and senile plaques.

Association of Aluminium with Neurofibrillary Tangle-Bearing Neurones

The application of imaging microanalytical techniques has demonstrated an association of aluminium with neurofibrillary tangle-bearing neurones, one of the major neuropathological features of Alzheimer's disease. Using scanning electron energy dispersive X-ray microanalysis, Perl and Brody [1980] found that aluminium was associated with the nuclear region in over 90% of neurofibrillary tangle-bearing neurones studied in formalin-fixed, silver-stained sections from Sommer's sector of the hippocampus in cases of senile dementia of the Alzheimer type. In contrast, aluminium was almost absent from adjacent non-tangle-bearing neurones and was rarely found in non-demented elderly patients. Recently, Kobayashi et al. [1987] have reported, using electron microprobe wavelength dispersive X-ray microanalysis, the presence of aluminium in neurofibrillary tangle-bearing neurones in an aluminium refiner who had Alzheimer's disease with Balint's syndrome, but failed to detect aluminium in neurofibrillary tangle-bearing neurones in a case of senile dementia of the Alzheimer type. The latter findings are particularly difficult to interpret due to the analysis of gold-coated, formalin-fixed, paraffin sections that had been stained with Weigert's iron haemotoxylin and eosin for analysis. Further evidence of an association between aluminium and neurofibrillary tangle-bearing neurones has been obtained in cases of parkinsonian dementia and amyotrophic lateral sclerosis on Guam. Perl et al. [1982], using scanning electron energy dispersive X-ray microanalysis, found that aluminium was associated with over 50% of the nuclear and perikaryal cytoplasmic sites examined in unstained, formalin-fixed sections of Sommer's sector of the hippocampus in cases of parkinsonian dementia and amyotrophic lateral sclerosis of Guam. In contrast, aluminium was only associated with 10% of the neuronal sites analysed in clinically normal Guamian patients. Confirmation of the association of aluminium with neurofibrillary tangle-bearing neurones in Sommer's sector of the hippocampus in Guamian cases has been obtained in unstained, formalin-fixed sections using electron microprobe wavelength dispersive X-ray microanalysis [Garruto et al., 1984, 1985, 1986]. Recently, Perl et al. [1986] have used laser microprobe mass analysis (LAMMA) to study the intracellular distribution of aluminium in Guamian cases in semithin resin-embedded sections and the highest level of aluminium was found to be associated with neurofibrillary tangles.

In the studies to date on the association of aluminium with neurofibrillary tangle bearing neurones only long-term formalin-fixed tissue has been used, this raises the possibility that redistribution of loss of loosely bound aluminium may have occurred. Furthermore, only Sommer's sector of the hippocampus has been studied. There is a clear need to confirm these findings in other brain regions in unfixed tissue sections and to study the intracellular distribution of aluminium in semithin resin-embedded sections where aluminium has been immobilized by low-temperature chelation.

Association of Aluminium and Silicon with Senile Plaques

Senile plaques are generally spherical structures, typically up to 200 μm in diameter which consist of dystrophic, degenerating and regenerating neuritic processes and glial cells (astrocytes and microglial cells) which in 'mature' plaques surround an extracellular core of amyloid. Terry and Wisniewski [1970] postulated that the sequence of senile plaque formation involved initially neuritic processes, followed at a later stage by the deposition of amyloid and finally loss of the neuritic processes to form a 'burned-out' plaque, which consist of only the amyloid core. This sequence of events no longer seems tenable since Miyakawa et al. [1982] have reported the presence of at least some amyloid fibrils in senile plaques studied in Alzheimer's disease, and Rudelli et al. [1984] have observed the presence of amyloid in white matter tracts. In Alzheimer's disease, Down's syndrome and the normal elderly individual, senile plaques occur predominantly in the neocortex, hippocampus and amygdala, although Rudelli et al. [1984] have observed senile plaques in the caudate nucleus, putamen, claustrum, hypothalamus and septal nuclei in Alzheimer's disease.

Initially, Duckett and Galle [1976], using electron microprobe wavelength dispersive X-ray microanalysis, found that aluminium was associated with senile plaques in formalin-fixed, cortical sections in Alzheimer's disease. In later brief reports these workers have observed, in a larger series of cases, that the highest concentration of aluminium was associated with lipofuscin granules in the rim of senile plaques and the cytoplasm of neurones, although it could sometimes be detected in senile plaque cores [Duckett and Galle, 1980; Duckett et al., 1985]. Unfortunately, these studies have not been fully documented, making the findings difficult to evaluate. Masters et al. [1985a] reported, using X-ray microanalysis, the presence of aluminium and silicon in isolated, intact senile plaque cores and also in the residue left after extraction of the isolated cores with formic acid. Stern et al. [1986] did not find prominent peaks for aluminium or silicon in the mass spectrum using LAMMA analysis of 0.5 μm sections of resin-embedded, isolated senile plaque cores. However, unless many serial, semithin sections were examined a focal concentration within the senile plaque core would not be detected using this approach.

High levels of silicon have been reported to be associated with both isolated and in situ senile plaque rims using both chemical and X-ray microanalysis respectively [Nikaido et al., 1972].

Our studies on senile plaque cores isolated from the cerebral cortex of cases of senile dementia of the Alzheimer type (fig. 1a) and Down's syndrome using scanning electron microprobe energy dispersive X-ray microanalysis, showed the presence of focal deposits of aluminium and silicon (fig. 1b) and revealed the co-localization of aluminium and silicon in the central region of the plaque core (fig. 1b) [Candy et al., 1984, 1985, 1986a]. High-resolution solid-state nuclear magnetic resonance studies of the isolated plaque core material demonstrated, for the first time, the presence of an aluminosilicate (fig. 2) [Candy et al., 1986a].

A potential pitfall in the analysis of isolated senile plaque core material is the presence of corpora amylacea which in contrast to senile plaque cores (fig. 1a) have a smooth topography (fig. 1c, e). Aluminium and sili-

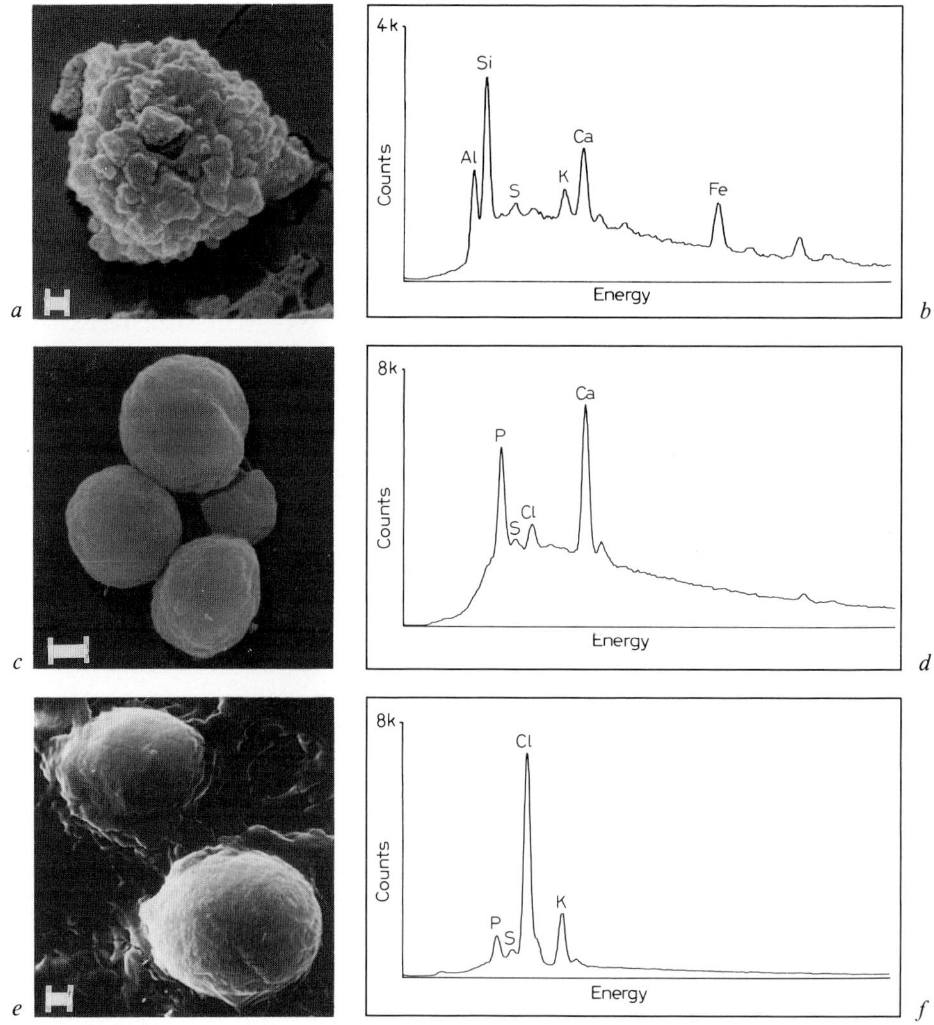

Fig. 1. Scanning electron micrographs and X-ray energy spectra of an isolated senile plaque core and isolated and in situ corpora amylacea. *(a)* Scanning electron micrograph and *(b)* X-ray energy spectrum of a senile plaque core isolated from the cerebral cortex of a patient with senile dementia of the Alzheimer type. Note rough surface and prominent peaks for aluminium and silicon. *(c)* Scanning electron micrograph and *(d)* X-ray energy spectrum of corpora amylacea isolated from the brain of a normal elderly individual.

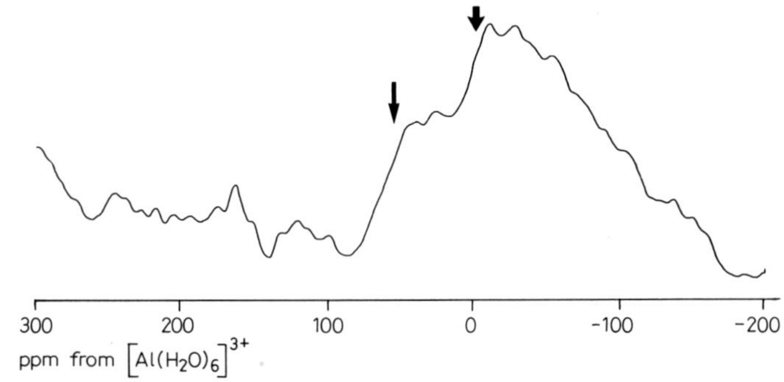

Fig. 2. ^{27}Al nuclear magnetic resonance spectrum obtained from senile plaque core material isolated from cases of senile dementia of the Alzheimer type. The spectrum shows the presence of both tetrahedral (long arrow) and octahedral aluminium (short arrow).

con were not detected in corpora amylacea isolated from normal brain (fig. 1d) or in situ in normal brain or in Alzheimer's disease (fig. 1f) using energy-dispersive X-ray microanalysis. Secondary ion mass spectrometric analysis of isolated corpora amylacea also did not reveal the presence of aluminium and silicon. These findings are in contrast to those of Nikaido et al. [1972] who reported high levels of silicon in isolated and in situ corpora amylacea.

We have demonstrated the focal co-localization of aluminium and silicon in the central region of in situ plaque cores using energy-dispersive X-ray microanalysis. These studies were carried out in silver stained and unstained cryostat sections of cerebral cortex from presenile and senile cases of Alzheimer's disease (fig. 3a, b), middle aged cases of Down's syn-

---

(e) Scanning electron micrograph and *(f)* X-ray energy spectrum of in situ corpora amylacea in an unfixed, unstained cryostat section of the basal forebrain of a patient with Alzheimer's disease. Note the similar smooth surface of isolated and in situ corpora amylacea and the lack of aluminium and silicon in both the isolated and in situ corpora amylacea. The prominent peaks for phosphorus and calcium in the isolated corpora amylacea are due to the isolation procedure. Calibration bars represent 2.5 μm.

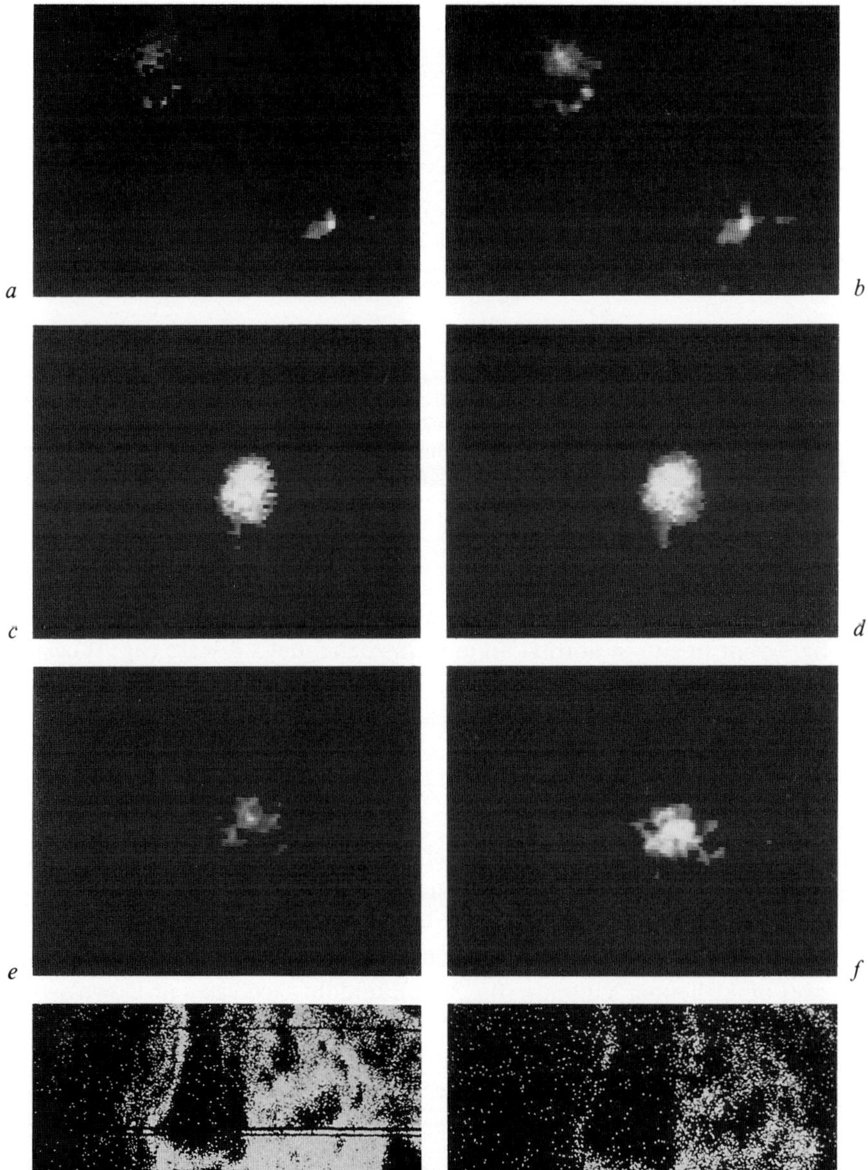

drome (fig. 3c, d) and normal elderly patients (fig. 3e, f) [Candy et al., 1985, 1986a; Edwardson et al., 1986]. An intrinsically more sensitive analytical technique for studying elemental distribution is scanning proton energy-dispersive X-ray microanalysis. Using this technique we have confirmed the presence of co-localized accumulations of aluminium and silicon in senile plaque cores and demonstrated the presence of iron within the plaque core [Candy et al., 1986b]. Secondary ion mass spectrometry is currently the most sensitive microanalytical technique for mapping elemental distribution and is especially suited to the localization of aluminium and silicon due to its high sensitivity for these elements and ability to make unequivocal elemental assignments for elements with a low mass. This technique has also shown the presence of focal, co-localized accumulations of aluminium and silicon in in situ senile plaque cores (fig. 3g, h).

Aluminium and Silicon Homeostasis

A potential transport mechanism for aluminium, in both the blood and brain, is the iron-carrying protein, transferrin. This protein may be responsible for the intracellular accumulation of aluminium and the presence of aluminium and iron in senile plaque cores. Transferrin has a relatively high affinity for aluminium [Cochran et al., 1984; Martin et al., 1987], transferrin receptors are present on the lumen of brain capillaries [Jefferies et al., 1984] and high levels of transferrin mRNA are present in rat brain [Levin et al., 1984]. We have found that aluminium-loaded transferrin can be as efficiently bound and internalized as iron-loaded transferrin by cultured IMR 32 neuroblastoma cells [Morris et al., 1987].

Furthermore, we have shown the presence of high levels of transferrin in human cerebral cortex and found that the level of transferrin and den-

---

Fig. 3. The presence of aluminium and silicon in senile plaque cores in silver-stained cryostat sections of cerebral cortex from Alzheimer's disease, Down's syndrome and normal elderly patients. *(a)* Aluminium and *(b)* silicon X-ray maps of a senile plaque core in senile dementia of the Alzheimer type. Calibration bar, 45 µm. *(c)* Aluminium and *(d)* silicon X-ray maps of a senile plaque core in Down's syndrome. Calibration bar, 8.1 µm. *(e)* Aluminium and *(f)* silicon X-ray maps in a normal elderly individual. Calibration bar, 18 µm. *(g)* Aluminium and *(h)* silicon high-resolution ion maps obtained using secondary ion mass spectrometry of a senile plaque core in senile dementia of the Alzheimer type. Calibration bar, 3 µm.

sity of transferrin receptors is unchanged in the cerebral cortex in senile cases of Alzheimer's disease [Morris et al., in press]. Autoradiographic studies in rat brain using ^{67}Ga as a potential marker for aluminium transport have shown that the high levels of ^{67}Ga are present in the cerebral cortex, hippocampus, septum and amygdala [Candy et al., 1987]. Interestingly, these areas also contain the highest densities of transferrin receptors [Candy et al., 1987] and show selective vulnerability in Alzheimer's disease.

In contrast to aluminium, much less is known about silicon. Silicon does not appear to be protein bound in plasma in either rat or man and may be present as magnesium orthosilicate [Berlyne et al., 1986]. Increased CSF levels have been reported in Alzheimer-type dementia [Hershey et al., 1984]. However, Austin et al. [1973] did not find an increased level of silicon in the brain in Alzheimer's disease, although the chemical method used would not have detected polymeric silicon.

Aluminium and the Pathogenesis of Neurofibrillary Tangles

Neurofibrillary tangles are a necessary but not exclusive neuropathological feature of Alzheimer's disease. In this disease the greatest density of neurofibrillary tangles is present in the neocortex, in contrast to their distribution in the elderly normal individual where it is restricted to archicortical areas [Perry, 1986]. The presence of neurofibrillary tangles in a number of diverse disease states, e.g. Alzheimer's disease, Down's syndrome, dementia pugilistica, parkinsonian dementia complex of Guam, Hallervorden-Spatz disease [Wisniewski et al., 1979], and the dissociation between their presence and the occurrence of senile plaques is difficult to explain on a genetic basis alone. The nature of the protein which constitutes the paired helical filament is at present controversial with conflicting claims that it is a low molecular weight protein [Masters et al., 1985a], high molecular weight neurofilament protein [Miller et al., 1986] or microtubule-associated protein tau [Kosik et al., 1986]. An important finding, that has emerged from these studies, is the ability of antibodies which recognize phosphorylated epitopes on neurofilaments and microtubule-associated protein tau to immunostain neurofibrillary tangles in neuronal soma in Alzheimer's disease [Sternberger et al., 1985; Miller et al., 1986; Grundke-Iqbal et al., 1986; Wood et al., 1986]. This aberrant phosphorylation may alter normal cytoskeletal interactions and impair the transport of, for example, neurofilaments. Aluminium-induced tangles are also immunostained by antibodies which recognize phosphorylated neurofilament epitopes [Troncoso et al., 1985; Munoz-Garcia et al., 1986]. The aberrant protein phosphorylation in Alzheimer's disease and that induced by aluminium may be related to impaired intracellular calcium homeostasis, with consequent activation of calcium-dependent protein kinases. Individual neuronal populations may differ in their susceptibility to impairments of calcium homeostasis depending on their calcium-buffering capacity [McBurney and Neering, 1987].

Impairments in calcium homeostasis have been reported in fibroblasts cultured from elderly normal individuals and patients with Alzheimer's disease [Peterson et al., 1986a, b] and Down's syndrome [Peterson et al., 1985]. The accumulation of aluminium within neurones in the parkinsonian-dementia complex of Guam has been suggested to be due to secondary hyperparathyroidism resulting from an environment rich in aluminium but low in calcium and magnesium [Garruto et al., 1984]. In relation to the occurrence of Alzheimer type neuropathological features in normal elderly individuals, it is interesting to note that a significant proportion of the elderly population is in negative calcium balance with decreased calcium absorption [Bullamore et al., 1970], due to reduced vitamin D hydroxylation. In rats calcium-magnesium deficiency has been shown to result in an increased level of aluminium in the CNS [Yase, 1980].

Aluminium can potentially interfere with a number of calcium- and magnesium-dependent processes. Binding of aluminium to calmodulin profoundly affects the flexibility of the protein and thereby modifies its interactions with other proteins [Siegel and Haug, 1983]. Aluminium can inhibit calcium-activated proteases [McDermott et al., unpubl.], this effect being probably due to the presence of a calmodulin-like domain in these proteins [Ohno et al., 1984]. Moreover, aluminium at subnanomolar concentrations has recently been shown to promote tubulin assembly and decrease its sensitivity to calcium-induced depolymerization [MacDonald et al., 1987]. Recently a plant neurotoxin (β-N-methylamino-*L*-alanine) has been linked with the Guam syndrome [Spencer et al., 1987]. This excitatory amino acid may act at N-methyl-*D*-aspartate receptors [Spencer et al., 1987] leading to an increase in cytoplasmic free calcium which may be neurotoxic [Choi, 1985]. Excitatory amino acids have previously been reported to induce the formation of paired helical filament-like structures in human neuronal cultures [De Boni and Crapper-McLachlan, 1985]. It is therefore conceivable that aluminium and an excitant amino acid may be involved in the pathogenesis of the Guam syndrome through a synergistic action on calcium homeostatic mechanisms.

There is also the possibility that aluminium may affect synthesis or degradation of a protein involved in paired helical filament formation. A marked nuclear accumulation of aluminium has been reported in neurofibrillary tangle-bearing neurones [Crapper-McLachlan and De Boni, 1980; Perl and Brody, 1980; Perl et al., 1982]. Aluminium has been shown to both stimulate and inhibit protein synthesis in different experimental paradigms [Exss and Summer, 1973; Miller and Levine, 1974; Magour and Maser, 1981]. Abnormal protein degradation could occur due to the inhibition of calcium-activated proteases by aluminium.

Aluminosilicates and the Pathogenesis of Senile Plaques

The spherical nature of senile plaques suggests that the stimulus for their formation may be centrally located within the plaque core. Recent advances in the characterization of the protein component of senile plaque core amyloid in Alzheimer's disease and Down's syndrome have shown that it is a low molecular weight, hydrophobic protein [Masters et al., 1985b], which is derived from a much larger precursor molecule that is coded for on chromosome 21 [Kang et al., 1987]. It seems unlikely that extracellular accumulation of a β-pleat sheat protein will lead to the pathological changes involved in senile plaque formation. Another candidate for these changes is the focal aluminosilicate deposits which occur within the senile plaque core. Aluminosilicates are highly active catalysts, for example, tetrahedral aluminium atoms are highly reactive even at levels of less than parts per million [Haag et al., 1984] and these species have been detected by nuclear resonance spectroscopy in isolated senile plaque core material (fig. 2) [Candy et al., 1986a].

The initial event in the formation of the aluminosilicate deposits is likely to be the polymerization of silicic acid either when its local concentration exceeds 100 parts per million or at much lower concentrations in the presence of aluminium [Iler, 1979]. Aluminium ions will then readily substitute into the silicate framework with the formation of aluminosilicates [Iler, 1979]. Iron which has also been found in senile plaques [Candy et al., 1987b] can also substitute into a silicate framework, although less readily than aluminium [Iler, 1979]. The amyloid protein may be secreted in response to the presence of the aluminosilicate by a mechanism similar to that responsible for silica-induced fibrogenesis [Aalto et al., 1980; Schmidt, 1984]. Secretion of a hydrophobic protein which results in the accumulation of fibrillary extracellular aggregates may thus serve to isolate the aluminosilicate.

References

Aalto, M.; Kulonen, E.; Ronnemaa, T.; Sundstrom, C.; Vilpo, J.: Liberation of a fibrogenic factor from human blood monocytes, ascites cells, cultured histiocytes and transformed mouse macrophages by treatment with SiO_2. Scand. J. clin. Lab. Invest. *40:* 311–318 (1980).

Alfrey, A.; LeGendre, G.; Kaehny, W.: The dialysis encephalopathy syndrome. Possible aluminium intoxication. New Engl. J. Med. *294:* 184–188 (1976).

Austin, J.; Rinehart, R.; Williamson, T.; Burcar, P.; Russ, K.; Nikaido, T.; LaFrance, M.: Studies in ageing of the brain. III. Silicon levels in postmortem tissues and body fluids. Prog. Brain Res. *40:* 485–495 (1973).

Berlyne, G.; Adler, A.; Ferran, N.; Bennett, S.; Holt, J.: Silicon metabolism. I. Some aspects of renal silicon handling in normal man. Nephron *43:* 5–9 (1986).

Bullamore, J.; Wilkinson, R.; Gallagher, J.; Nordin, B.; Marshall, D.: Effect of age on calcium absorption. Lancet *ii:* 535–537 (1970).

Candy, J.; Edwardson, J.; Faircloth, R.; Keith, A.; Morris, C.; Pullen, R.: Gallium-67 as a potential marker for aluminium transport in rat brain. J. Physiol. *391:* 34P (1987).

Candy, J.; Edwardson, J.; Klinowski, J.; Oakley, A.; Perry, E.; Perry, R.: Co-localisation of aluminium and silicon in senile plaques: implications for the neurochemical pathology of Alzheimer's disease; in Traber, Gispen, Senile dementia of Alzheimer type, pp. 183–197 (Springer, Heidelberg 1985).

Candy, J.; Oakley, A.; Atack, J.; Perry, R.; Perry, E.; Edwardson, J.: New observations on the nature of senile plaque cores; in Proc. 5th Meet. Eur. Soc. Neurochem., Budapest 1984. Vizi, Magyar. Regulation of transmitter function: Basic and clinical aspects, pp. 301-304 (Akadémiai Kiadó, Budapest 1984).

Candy, J.; Oakley, A.; Klinowski, J.; Carpenter, T.; Perry, R.; Atack, J.; Perry, E.; Blessed, G.; Fairbairn, A.; Edwardson, J.: Aluminosilicates and senile plaque formation in Alzheimer's disease. Lancet *i:* 354–357 (1986a).

Candy, J.; Oakley, A.; Watt, F.; Grime, G.; Klinowski, J.; Perry, R.; Edwardson, J.: A role for aluminium, silicon and iron in the genesis of senile plaques; in Colloque Inserm-Eurage, Modern trends in aging research, vol. 147, pp. 443–450 (Libbey, London 1986b).

Choi, D.: Glutamate neurotoxicity in cortical cell culture is calcium dependent. Brain Res. *58:* 293–297 (1985).

Cochran, M.; Coates, J.; Neoh, S.: The competitive equilibrium between aluminium and ferric ions for the binding sites of transferrin. FEBS Letts. *176:* 129–132 (1984).

Crapper, D.: Functional consequences of neurofibrillary degeneration; in Terry, Gershon, Neurobiology of aging, pp. 405–432 (Raven Press, New York 1976).

Crapper, D.; Krishman, S.; Dalton, A.: Brain aluminium distribution in Alzheimer's disease and experimental neurofibrillary degeneration. Science *180:* 511–513 (1973).

Crapper, D.; Krishman, S.; Quittkat, S.: Aluminium, neurofibrillary degeneration and Alzheimer's disease. Brain *99:* 67–80 (1976).

Crapper-McLachlan, D.; De Boni, U.: Aluminium in human brain disease: an overview. Neurotoxicity *1:* 3–16 (1980).

De Boni, U.; Crapper- MacLachlan, D.: Controlled induction of paired helical filaments of the Alzheimer type in cultured human neurons, by glutamate and aspartate. J. neurol. Sci. *68:* 105–118 (1985).

Duckett, S.; Galle, P.: Mise en évidence de l'aluminium dans les plaques séniles de la maladie d'Alzheimer: étude à la microsonde de Castaing. C.r. hebd. Séanc. Acad. Sci., Paris *282:* 393–395 (1976).

Duckett, S.; Galle, P.: Electron microscope-microprobe studies of aluminium in the brains of cases of Alzheimer's disease and ageing patients. J. Neuropath. exp. Neurol. *39:* 350 (1980).

Duckett, S.; Galle, P.; Fiori, C.: Electron probe microanalysis of normal and pathological neural tissue with wavelength dispersive X-ray spectrometry; in Gabay, Harris, Ho, Metal ions in neurology and psychiatry, pp. 367–396 (Liss, New York 1985).

Edwardson, J.; Klinowski, J.; Oakley, A.; Perry, R.; Candy, J.: Aluminosilicates and the ageing brain: implications for the pathogenesis of Alzheimer's disease; in Ciba Fdn Symp. 121, pp. 160–179 (Wiley, Chichester 1986).

Exss, R.; Summer, G.: Basic proteins in neurons containing fibrillary deposits. Brain Res. *49:* 151–154 (1973).

Garruto, R.; Fukatsu, R.; Yanagihara, R.; Gajdusek, D.; Hook, G.; Fiori, C.: Imaging of calcium and aluminium in neurofibrillary tangle-bearing neurons in parkinsonian-dementia of Guam. Proc. natn. Acad. Sci. USA *81:* 1875–1879 (1984).

Garruto, R.; Swyt, C.; Fiori, C.; Yanagihara, R.; Gajdusek, D.: Intraneuronal deposition of calcium and aluminium in amyotrophic lateral sclerosis of Guam. Lancet *ii:* 1353 (1985).

Garruto, R.; Swyt, C.; Yanagihara, R.; Fiori, C.; Gajdusek, D.: Intraneuronal co-localisation of silicon with calcium and aluminium in amyotrophic lateral sclerosis and parkinsonism with dementia of Guam. New Engl. J. Med. *315:* 711–712 (1986).

Grundke-Iqbal, I.; Iqbal, K.; Tung, Y.-C.; Quinlan, M.; Wisniewski, H.; Binder, L.: Abnormal phosphorylation of the microtubule-associated protein τ (tau) in Alzheimer cytoskeletal pathology. Proc. natn. Acad. Sci. USA *83:* 4913–4917 (1986).

Haag, W.; Lago, R.; Weisz, P.: The active site of acidic aluminosilicate catalysis. Nature, Lond. *309:* 589–591 (1984).

Hershey, L.; Hershey, C.; Varnes, A.: CSF silicon in dementia: A prospective study. Neurology *34:* 1197–1201 (1984).

Iler, R.: The chemistry of silica. Solubility, polymerisation, colloid and surface properties, and biochemistry, pp. 1–866 (Wiley, New York 1979).

Jefferies, W.; Brandon, M.; Hunt, S.; Williams, A.; Gatter, K.; Mason, D.: Transferrin receptor on endothelium of brain capillaries. Nature, Lond, *312:* 162–163 (1984).

Kang, J.; Lemaire, H.-G.; Unterbeck, A.; Salbaum, J.; Masters, C.; Grzeschik, K.-H.; Multhaup, G.; Beyreuther, K.; Muller-Hill, B.: The precursor of Alzheimer's disease amyloid A4 protein resembles a cell-surface receptor. Nature, Lond. *325:* 723–736 (1987).

Kidd. M.: Paired helical filaments in electron microscopy in Alzheimer's disease. Nature, Lond. *197:* 192–193 (1963).

Klatzo, I.; Wisniewski, H.; Streicher, E.: Experimental production of neurofibrillary degeneration. I. Light microscopic observations. J. Neuropath. exp. Neurol. *24:* 187–199 (1965).

Kobayashi, S.; Hirota, N.; Saito, K.; Utsuyama, M.: Aluminium accumulation in tangle-bearing neurons of Alzheimer's disease with Balint's syndrome in a long-term aluminium refiner. Acta neuropath. *74:* 47–52 (1987).

Kosik, K.; Joachim, C.; Selkoe, D.: Microtubule-associated protein tau (τ) is a major antigenic component of paired helical filaments in Alzheimer's disease. Proc. natn. Acad. Sci. USA *83:* 4044–4048 (1986).

Levin, M.; Tuil, D.; Uzan, G.; Dreyfus, J.-C.; Kahn, A.: Expression of the transferrin gene during development of non-hepatic tissues. Biochem. biophys. Res. Commun. *122:* 212–217 (1984).

MacDonald, T.; Humphreys, W.; Martin, J.: Promotion of tubulin assembly by aluminium ion in vitro. Science *236:* 183–186 (1987).

Magour, S.; Maser, H.: Effect of acute treatment with aluminium chloride on brain microsomal protein synthesis of immature rats. Biochem. Soc. Trans. *8:* 100–101 (1981).

Markesbery, W.; Ehmann, W.; Hossain, B.; Alauddin, M.; Goodin, M.: Instrumental neutron activation analysis of brain aluminium in Alzheimer disease and aging. Ann. Neurol. *10:* 511–516 (1981).

Martin, R.; Savory, J.; Brown, S.; Bertholf, R.; Wills, M.: Transferrin binding of Al^{3+} and Fe^{3+}. Clin. Chem. *33:* 405–407 (1987).

Masters, C.; Multhaup, G.; Simms, G.; Pottgiesser, J.; Martins, R.; Beyreuther, K.: Neuronal origin of a cerebral amyloid: neurofibrillary tangles of Alzheimer's disease contain the same protein as the amyloid of plaque cores and blood vessels. Eur. molec. Biol. Org. J. *4:* 2757–2763 (1985a).

Masters, C.; Simms, G.; Weinman, N.; Multhaup, G.; McDonald, B.; Beyreuther, K.: Amyloid plaque core protein in Alzheimer's disease and Down's syndrome. Proc. natn. Acad. Sci. USA *82:* 4245–4249 (1985b).

McBurney, R.; Neering, I.: Neuronal calcium homeostasis. Trends Neurosci. *10:* 164–169 (1987).

McDermott, J.; Smith, A.; Iqbal, K.; Wisniewski, H.: Brain aluminium in aging and Alzheimer's disease. Neurology *29:* 809–814 (1979).

McLaughlin, A.; Kazontzis, G.; King, E.; Teare, D.; Porter, R.; Owen, R.: Pulmonary fibrosis and encephalopathy associated with the inhalation of aluminium dust. Br. J. ind. Med. *19:* 253 (1962).

Miller, C.; Brion, J.-P.; Calvert, R.; Chin, T.; Eagles, P.; Downes, M.; Flament-Durand, J.; Haugh, M.; Kahn, J.; Probst, A.; Ulrich, J.; Anderton, B.: Alzheimer's paired helical filaments share epitopes with neurofilament side arms. Eur. molec. Biol. Org. J. *5:* 269–276 (1986).

Miller, C.; Levine, E.: Effects of aluminium salts on cultured neuroblastoma cells. J. Neurochem. *22:* 751–758 (1974).

Morris, C.; Candy, J.; Court, J.; Whitford, C.; Edwardson, J.: The role of transferrin in the uptake of aluminium and manganese by the IMR 32 neuroblastoma cell line. Biochem. Soc. Trans. *15:* 498 (1987).

Morris, C.; Court, J.; Moshtaghie, A.; Skillen, A.; Candy, J.; Perry, R.; Edwardson, J.: Transferrin and transferrin receptors in normal brain and in Alzheimer's disease. Biochem. Soc. Trans. (in press).

Miyakawa, T.; Shimoji, A.; Kuramoto, R.; Higuchi, Y.: The relationship between senile plaques and cerebral blood vessels in Alzheimer's disease and senile dementia. Virchows Arch. *40:* 121–129 (1982).

Munoz-Garcia,D.; Pendelbury, W.; Kessler, J.; Perl, D.: An immunocytochemical comparison of cytoskeletal proteins in aluminium-induced and Alzheimer-type neurofibrillary tangles. Acta neuropath. *70:* 243–248 (1986).

Nikaido, T.; Austin, J.; Trueb, L.; Rinehart, R.: Studies in ageing of the brain. II. Microchemical analyses of the nervous system in Alzheimer patients. Archs Neurol. *27:* 549–554 (1972).

Ohno, S.; Emori, Y.; Imajoh, S.; Kawasaki, H.; Kisaragi, M.; Suzuki, K.: Evolutionary origin of a calcium-dependent protease by fusion of genes for a thiol protease and a calcium-binding protein? Nature, Lond. *312:* 566–570 (1984).

Perl, D.; Brody, A.: Alzheimer's disease: X-ray spectrometric evidence of aluminium accumulation in neurofibrillary tangle-bearing neurones. Science *208:* 297–299 (1980).

Perl, D.; Gajdusek, D.; Garruto, R.; Yanagihara, R.; Gibbs, C.: Intraneuronal aluminium accumulation in amyotrophic lateral sclerosis and parkinsonism dementia of Guam. Science *217:* 1053–1055 (1982).

Perl, D.; Munoz-Garcia, D.; Good, P.; Pendlebury, W.: Laser microprobe mass analyzer (LAMMA) – a new approach to the study of the association of aluminium and neurofibrillary tangle formation; in Fisher, Lachman, Harris, Advances in behavioural biology, vol 29: Alzheimer's and Parkinson's disease. strategies for research and development (Plenum Press, New York 1986).

Perry, R.: Recent advances in neuropathology. Br. med. Bull. *42:* 34–41 (1986).

Peterson, C.; Gibson, G.; Blass, J.: Altered calcium uptake in cultured skin fibroblasts from patients with Alzheimer's disease. New Engl. J. Med. *312:* 1063–1065 (1985).

Petersen, C.; Goldman, J.: Alterations in calcium content and biochemical processes in cultured skin fibroblasts from aged and Alzheimer donors. Proc. natn. Acad. Sci. USA *83:* 2758–2762 (1986a).

Peterson, C.; Ratan, R.; Shelanski, M.; Goldman, J.: Cytoplasmic free calcium and cell spreading decrease in fibroblasts from aged and Alzheimer donors. Proc. natn. Acad. Sci. USA *83:* 7999–8001 (1986b).

Rudelli, R.; Ambler, M.; Wisniewski, H.: Morphology and distribution of Alzheimer neuritic (senile) and amyloid plaques in striatum and diencephalon. Acta neuropath. *64:* 273–281 (1984).

Schmidt, J.; Oliver, C.; Lepe-Zuniga, J.; Green, I.; Gery, I.: Silica-stimulated monocytes release fibroblast proliferation factors identical to interleukin-1. A potential role for interleukin-1 in the pathogenesis of silicosis. J. clin. Invest. *73:* 1462–1472 (1984).

Siegel, N.; Haugh, A.: Aluminium interaction with calmodulin. Biochim. biophys. Acta *744:* 36–45 (1983).

Spencer, P.; Nunn, P.; Hugon, J.; Ludolph, A.; Ross, S.; Roy, D.; Robertson, R.: Guam amyotrophic lateral sclerosis – parkinsonism – dementia linked to a plant excitant neurotoxin. Science *237:* 517–522 (1987).

Stern, A.; Perl, D.; Munoz-Garcia, D.; Good, R.; Abraham, C.; Selkoe, D.: Investigation of silicon and aluminium content in isolated senile plaque cores by laser microprobe mass analysis (LAMMA). J. Neuropath. exp. Neurol. *45:* 361 (1986).

Sternberger, N.; Sternberger, L.; Ulrich, J.: Aberrant neurofilament phosphorylation in Alzheimer's disease. Proc. natn. Acad. Sci. USA *82:* 4274–4276 (1985).

Terry, R.; Pena, C.: Experimental production of neurofibrillary degeneration. II. Electron microscopy, phosphatate chemistry and electron probe analysis. J. Neuropath. exp. Neurol. *24:* 200–210 (1965).

Terry, R.; Wisniewski, H.: The ultrastructure of the neurofibrillary tangle and the senile plaque; in Wolstenholme, O'Connor, Alzheimer's disease and related conditions, pp. 145–168 (Churchill, London 1970).

Trapp, G.; Miner, G.; Zimmerman, R.; Mastri, A.; Heston, L.: Aluminium levels in brain in Alzheimer's disease. Biol. Psychiat. *13:* 709–718 (1978).

Traub, R.; Rains, T.; Garruto, R.; Gajdusek, D.; Gibbs, C.: Brain destruction alone does not elevate brain aluminium. Neurology *31:* 986–990 (1981).

Troncoso, J.; Sternberger, L.; Sternberger, N.; Hoffman, P.; Price, D.: Immunocytochemical studies of neurofilament antigens in the neurofibrillary pathology induced by aluminium. J. Neuropath. exp. Neurol. *44:* 332 (1985).

Ward, N.; Mason, J.: Neutron activation analysis techniques for identifying elemental status in Alzheimer's disease; in Heydorn, Modern trends in activation analysis, pp. 925–934 (Isotope Division, Riso National Laboratory, Roskilde 1986).

Wisniewski, K.; Jervis, G.; Moretz, R.; Wisniewski, H.: Alzheimer neurofibrillary tangles in diseases other than senile and presenile dementia. Ann. Neurol. *5:* 288–294 (1979).

Wood, J.; Mirro, S.; Pollock, N.; Binder, L.: Neurofibrillary tangles of Alzheimer's disease share antigenic determinants with the axonal microtubule-associated protein tau (τ). Proc. natn. Acad. Sci. USA *83:* 4040–4043 (1986).

Yase, Y.: The role of aluminium in CNS degeneration with interaction of calcium; in Liss, Aluminium neurotoxicity, pp. 101–109 (Pathotox, Illinois 1980).

Dr. J.M. Candy, MRC Neuroendocrinology Unit, Newcastle General Hospital, Westgate Road, Newcastle upon Tyne NE4 6BE (UK)

Interdiscipl. Topics Geront., vol. 25, pp. 156–172 (Karger, Basel 1988)

Differential Pattern of Senile Plaque Neurites Immunoreactivity Related to the Presence or Absence of Neurofibrillary Tangles in the Neocortex

A. Probst[a], *C. Lautenschlager*[a], *J.P. Brion*[b], *B.H. Anderton*[c], *J. Ulrich*[a]

[a] Department of Pathology, Division of Neuropathology, University of Basel, Switzerland; [b] Laboratoire d'Anatomie Pathologique et de Microscopie Electronique, Université Libre de Bruxelles, Belgique, and [c] Department of Immunology, St. George's Hospital Medical School, London, UK

Introduction

Senile plaques (SP) and neurofibrillary tangles (NFT) are characteristics of the morphology of Alzheimer's brain. Demonstrated by histochemical stain (thioflavine S), silver impregnation and immunocytochemical methods, NFT occur in perikarya of neurons in different regions of the brain, including the hippocampus (mainly the CA 1 sector), amygdala, neocortex, locus coeruleus, dorsal raphe and nucleus basalis of Meynert. Ultrastructurally, NFT are large nonmembrane bound masses of filaments in the perinuclear cytoplasm of neurons. Most of these NFT have the appearance of filaments (approximately 10–13 nm in diameter) helically twisted round each other in pairs with a half periodicity of about 80 nm [Kidd, 1963; Wisniewski and Terry, 1976]. These twisted filaments are now referred to as paired helical filaments (PHF). SP are roughly spherical areas of altered neuropil measuring 10–150 μm in diameter, mainly occurring in the hippocampus, cerebral cortex and amygdaloid nucleus. The main constituents of the SP are extracellular deposits of amyloid, altered neurites and 'reactive cells' (microglial cells and astrocytes) [Terry and Wisniewski, 1970].

Despite considerable progress made during the last decade in understanding the molecular pathology of NFT and in particular of the amyloid fibrils of the SP, the pathogenesis of these two lesions remains largely unknown and the problem of their relationship unsolved. In particular, although PHF have long been documented in abnormal neurites of SP [Wisniewski et al., 1981] we do not know if this is of pathogenetic relevance for the formation of SP. As there have been no projection studies it remains unclear whether these abnormal filaments in plaque neurites are transported from NFT-bearing perikarya or are formed locally within distal axons or dendrites of neurons participating in the SP. It was stated that the number of PHF-filled neurites in the plaque is proportional to the number of nerve cells containing NFT [Wisniewski et al., 1981]. This relationship suggests that PHF in plaque neurites are dependent on the presence of tangles in neuronal cell bodies situated in the vicinity of plaques. It suggests furthermore that neurites in the plaques may fail to contain PHF in the absence of NFT in local neurons. If this proves to be true, we would have to consider PHF as a nonobligate feature of SP only reflecting the presence of NFT in local neurons innervating SP by chance. In other terms, development of SP would not imply the presence of PHF, but both processes could occur independently in the same brain.

Absence of neocortical tangles in the presence of numerous SP has been recently documented in a significant minority (30%) of a large series of demented patients aged more than 74 [Terry et al., 1987]. Patients without neocortical tangles did not significantly differ from patients with neocortical tangles after the degree of dementia, Blessed score or brain weight. A reverse situation is observed in a small number of apparently unrelated conditions including dementia pugilistica, Guam dementia, postencephalitic Parkinson's disease and subacute sclerosing panencephalitis. Here, NFT alone occur without amyloid deposition and SP formation.

The present study aims at getting more information on the relationship between NFT and SP, especially between NFT and the presence of altered structural proteins in SP neurites. For this purpose, we used Gallyas silver impregnation method [Gallyas, 1971] which selectively stains PHF [Braak et al., 1986] and a panel of antibodies directed against normal and pathological cytoskeleton proteins. We were particularly interested to know if the antigenic properties of SP neurites are modified according to the presence or absence of NFT in the same areas of the neocortex. Our results confirm Terry's [1987] observation that SP may occur in the total absence of NFT in the neocortex in a sizable proportion of SDAT patients. They furthermore demonstrate that PHF epitopes are not an essential feature of SP neurites as they are consistently lacking in SP in the absence of NFT.

Materials and Methods

Selection of Cases

Brains used for this study were obtained at autopsy from 10 old patients suffering from SDAT, from 2 intellectually normal old patients and 2 normal young patients. Sex, age, main clinical diagnosis and neuropathology are listed in table I.

These 10 SDAT brains were selected from a large file of demented patients autopsied and routinely examined in our institution. Two groups of patients were made: group I (5 patients) with tangles and SP in the neocortex, and group II (5 patients) with SP but no tangles in at least three regions of the neocortex. Demented patients were referred to autopsy from gerontological and psychiatric wards. The degree of dementia had been assessed by the 'mini mental state' protocol of Folstein et al. [1975]. Beside clinically established progressive dementia, the criterium for ATD was histological evidence of a large number of SP (more than 10 or 15/mm^2), depending on the age of the patients [Khachaturian, 1985], in several neocortical areas, hippocampus and amygdala. The 5 cases of group II have been included in an earlier study by Ulrich and Stähelin [1984].

Tissue Processing

For histology, paraffin sections were cut at 4 μm through the amygdala, the hippocampal formation and the adjacent entorhinal cortex, the gyrus rectus (Brodmann's area 11), the superior frontal gyrus (area 9), the temporal gyrus (area 22), the supramarginal

Table I. Sources of brain tissue

Case No.	Sex	Age at death	Clinical diagnosis	Neuropathology
1	F	72	severe dementia	SDAT
2	M	83	severe dementia	SDAT
3	M	86	dementia	SDAT
4	M	78	severe dementia	SDAT
5	F	90	dementia	SDAT
6	M	88	dementia	SDAT, deep lacunar infarcts
7	F	82	severe dementia	SDAT, deep lacunar infarcts
8	M	84	dementia	SDAT
9	M	88	dementia	SDAT
10	F	95	dementia	SDAT
11	M	84	intellectually normal (IN)	no lesions (NL)
12	M	78	IN	NL
13	M	46	IN	NL
14	F	42	IN	NL

Patients 1–5 belong to group I (both SP and tangles in the neocortex) and patients 6–10 to group II (SP in the neocortex without tangles).

gyrus (area 40), the dorsal occipital cortex (area 19) and the primary visual cortex (area 17). The sections (4 µm) were deparaffinized and stained with HE, PAS, Holmes silver impregnation for axons combined with luxol fast blue for myelin sheets (HL) and the fluorochrome thioflavine S [Schwartz, 1970] for the demonstration of NFT and amyloid cores of the SP. Additional silver staining was performed according to the method of Gallyas [1971]: briefly, sections were deparaffinized in xylene/alcohol, rinsed in distilled water and treated for 30 min with 5% periodic acid. After a brief rinse they were impregnated in alkaline silver iodide complex solution for 10 min. Sections were then left in physical developer for 30 min and transferred into a 0.5% solution of acetic acid to terminate development.

Immunocytochemistry was performed on 3–4 different neocortical areas and in the hippocampus in all cases. A panel of antibodies against cytoskeletal proteins was used, whose predominant specificities are listed in table II. Sections were deparaffinized, rinsed in phosphate-buffered saline (PBS) and endogenous peroxidase activity was stopped with 0.3% hydrogen peroxide in methanol (30 min). Subsequently, sections were treated with normal horse (for monoclonal antibodies) or normal goat serum (polyclonal antibodies). Incubations were performed overnight at +4 °C in diluted solutions of antibodies, followed by thorough washing in PBS, treatment with specific biotinylated anti-mouse or anti-rabbit immunoglobulin (30 min) avidin-biotin-peroxidase complex (Vectastain) (30 min) and another washing in PBS. Negative controls were made by replacing first antibodies with normal rabbit or mouse serum. Peroxidase activity was developed with di-

Table II

Silver stain and antibodies used in the study	Predominant specificity
Gallyas silver impregnation	PHF in situ (in NFT, 'neuropil threads' and SP neurites) [Gallyas, 1971; Braak et al., 1986] PHF and tau polypeptide in SDS polyacrylamide gel [Iqbal et al., 1987] normal neurofilaments are not stained
Polyclonal antibodies	
Anti-PHF	NFT in situ (neuronal perikarya, SP neurites) and SDS-isolated [Brion et al., 1985] partially MAP-tau
Anti-MAP-tau	MAP-tau NFT in situ [Brion et al., 1985]
Monoclonal antibodies	
BF 10	NFT in situ (neuronal perikarya and SP neurites) phosphorylated epitopes on human neurofilaments (major antigenic determinants: 155,000) [Anderton et al., 1982]
RT 97	like BF 10 (major antigenic determinant: 210,000)

aminobenzidine /H_2O_2 solution. Series of ten consecutive sections (4 μm each) were made from two neocortical areas (supramarginal gyrus and dorsal occipital cortex) from the 5 patients of group II (only NP in neocortex without tangles). Consecutive sections of the series were stained with Gallyas, a polyclonal antibody against isolated PHF and an anti-MAP-tau serum.

Results

Neuropil Threads

In Gallyas-impregnated sections of iso- and allocortical areas of group I SDAT patients, thread-like argyrophilic fibers (threads) similar to those described by Braak et al. [1986] were found in various amounts throughout the neuropil next to intensely stained NFT in the soma of cortical nerve cells (fig. 1, 2, 4a). Threads generally appeared as curvilinear or wave-shaped profiles of regular caliber measuring 1–2 μm in thickness. Short spindle-shaped argyrophilic profiles (about 2 μm thick) were less frequently encountered in the neuropil. In general, the density of neuropil threads was proportional to the density of NFT in the same cortical area. In the entorhinal cortex, where NFT were frequently arranged in laminar rows, threads were concentrated in the involved layers (fig. 1). In cortical areas where only few scattered NFT were present, only few threads could be found which were generally concentrated in the immediate surroundings of tangles.

Threads generally extended through the whole thickness of the cortex, although their density was much less in the sixth cortical layer. Only occasional threads could be found in the subcortical white matter. In the hippocampus, no thread-like profiles were found in the alveus, even in the presence of numerous threads and tangles in the corresponding allocortex including the stratum oriens.

On adjacent sections immunostained with the anti-PHF and the anti-MAP-tau, numerous immunoreactive thread-like processes could also be seen throughout the allo- and neocortical neuropil (fig. 3b) next to intensely labeled NFT. These processes had the same pattern of distribution and identical morphological characteristics as those found in Gallyas silver preparations although they appeared to be somewhat less numerous than silver positive profiles.

In group II SDAT patients (without neocortical NFT) there were, however, numerous Gallyas-positive and immunoreactive threads and NFT in

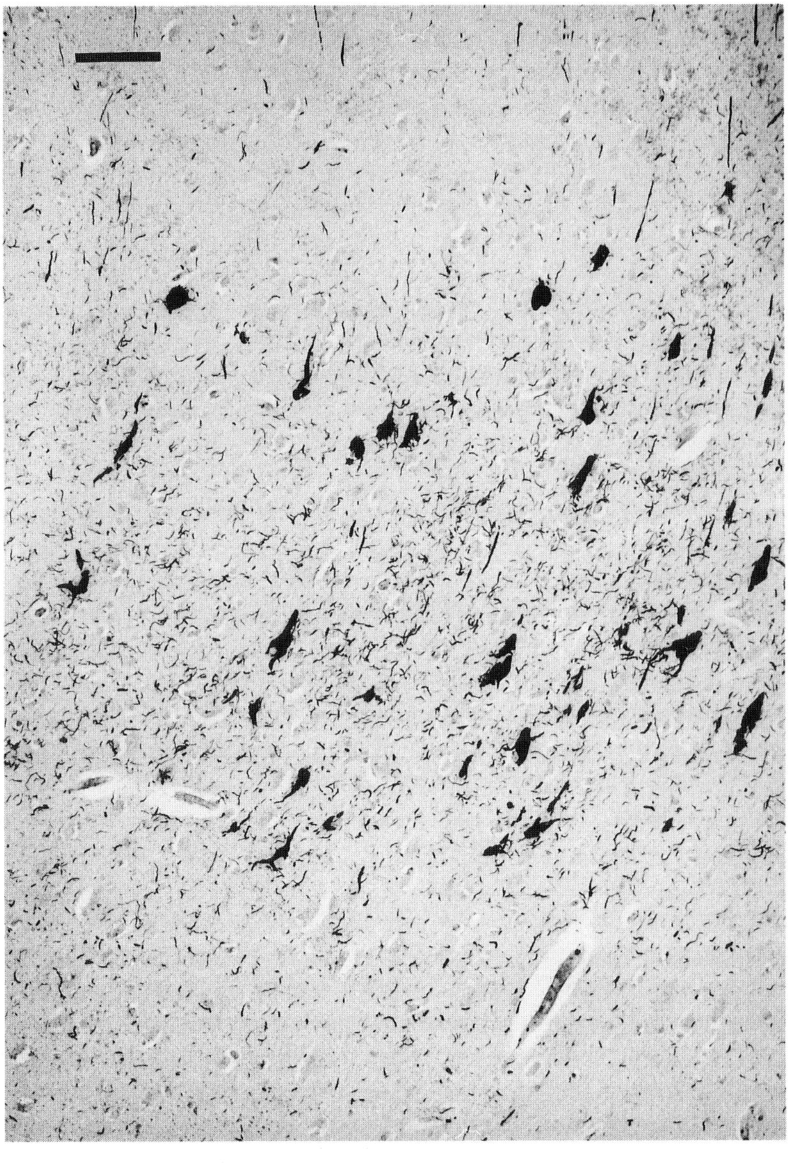

Fig. 1. Gallyas impregnation of neuropil threads and NFT within the entorhinal cortex. Note increased densities of threads in the layers containing the NFT suggesting that threads correspond to dendritic arborization of involved neurons. 4-μm thick paraffin sections without counterstain. Bar = 100 μm.

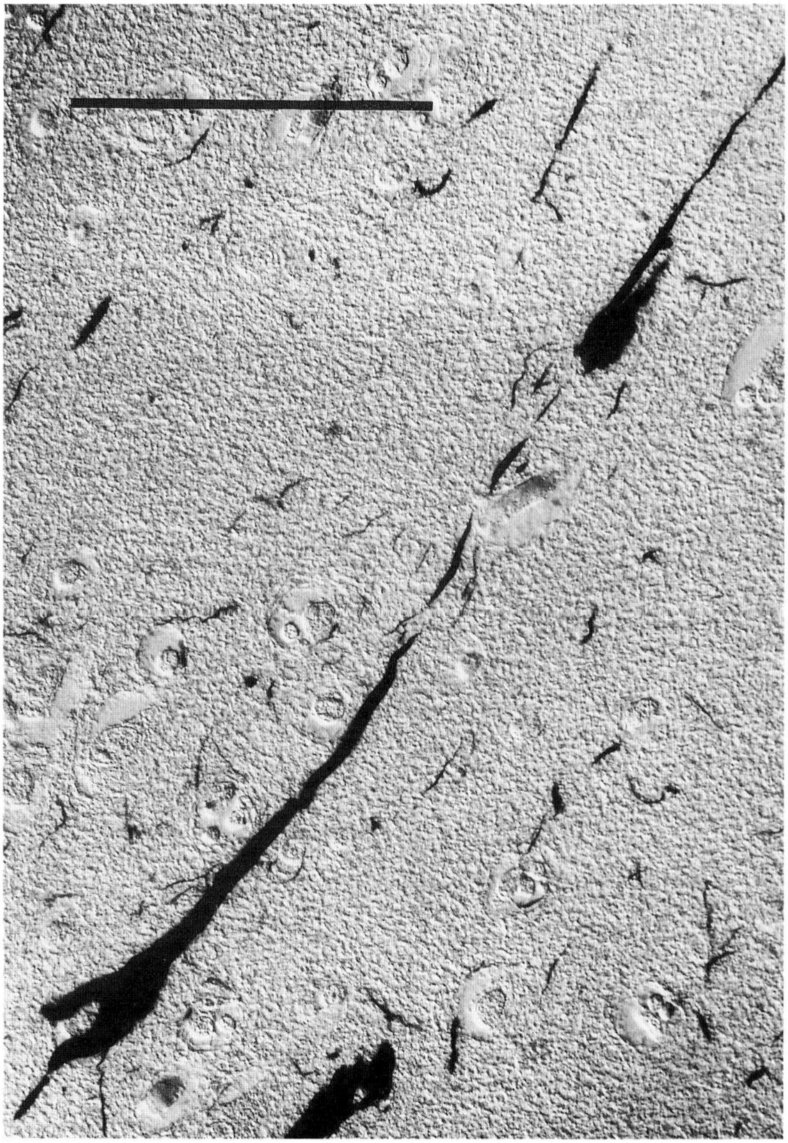

Fig. 2. Gallyas impregnation of a pyramidal cell in entorhinal cortex. Pathologic material is stained in the perikaryon and cell processes, mainly in the apical dendrite. 4-μm thick paraffin section. Nomarski interference contrast optics. No counterstain. Bar = 100 μm.

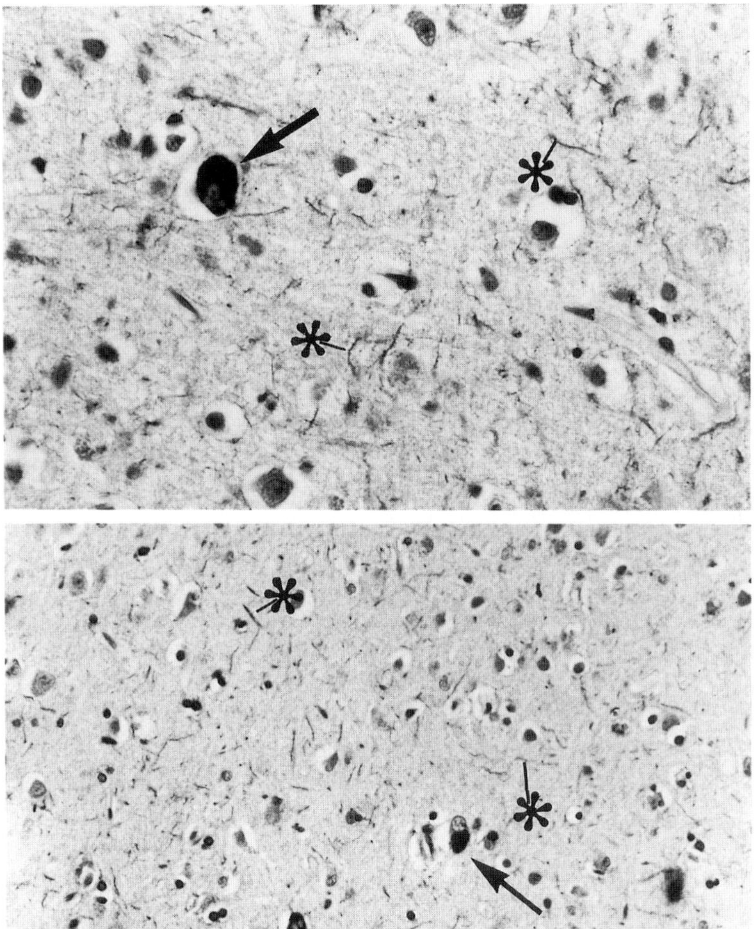

a

b

Fig. 3. a Anti-MAP-tau labeling of the neocortex in a SDAT patient of group I. Arrow indicates a NFT and asterisks tau-positive threads in the neuropil. 4-μm thick paraffin section counterstained with hematoxylin. ×438. *b* Anti-PHF labeling of the neocortex in a SDAT case of group I. Arrows indicate NFT and asterisks anti-PHF-positive threads in the cortical neuropil. 4-μm thick paraffin section counterstained with hematoxylin. ×219.

the examined allocortical areas and the amygdala. However, NFT were totally lacking in the selected neocortical regions and there were no Gallyas-positive or immunoreactive threads in the neuropil.

Neocortical Senile Plaques

The staining results in SDAT group I and II patients are summarized in table III. There were no consistent differences in morphology and especially in the number of amyloid-rich SP in the two patient groups. In group I patients, SP were easily recognized in Gallyas iodide stain as aggregates of tortuous argyrophilic processes surrounding a pale nonstained center, corresponding to the amyloid core (fig. 4a). Impregnated processes of the SP neuritic corona were generally more conspicuous than the neurites found outside of the plaque area. Variable numbers of pathologic neurites in SP were also labelled with all monoclonal and polyclonal antibodies of the pannel (fig. 5a). Degenerating neurites were, however, more conspicuously stained with the anti-PHF (fig. 5a) than with the other antibodies. The central amyloid core in SP of the 'classic' or 'mature' type remained consistently unstained both by the Gallyas method and by immunocytochemistry.

Table III. Gallyas impregnation and immunohistochemical reactions of SP neurites in the presence and absence of cortical NFT

	Group I, patients with NFT in neocortex	Group II, patients with NFT in neocortex
Gallyas	N (numerous threads in neuropil)	– (no threads in neuropil)
Polyclonal anti-PHF	n	–
Polyclonal anti-MAP-tau	n	–
Monoclonal RT 97	n	n
Monoclonal BF 10	n	n

N = most neurites impregnated; n = variable number of reactive neurites; – = no impregnated or immunoreactive neurites.

Fig. 4. a Gallyas impregnation of the neocortex in a SDAT patient of group I, showing numerous silver-impregnated neurites in a senile plaque and neuropil threads in the surrounding neuropil. 4-μm thick paraffin section. No counterstain. ×438. *b* Gallyas

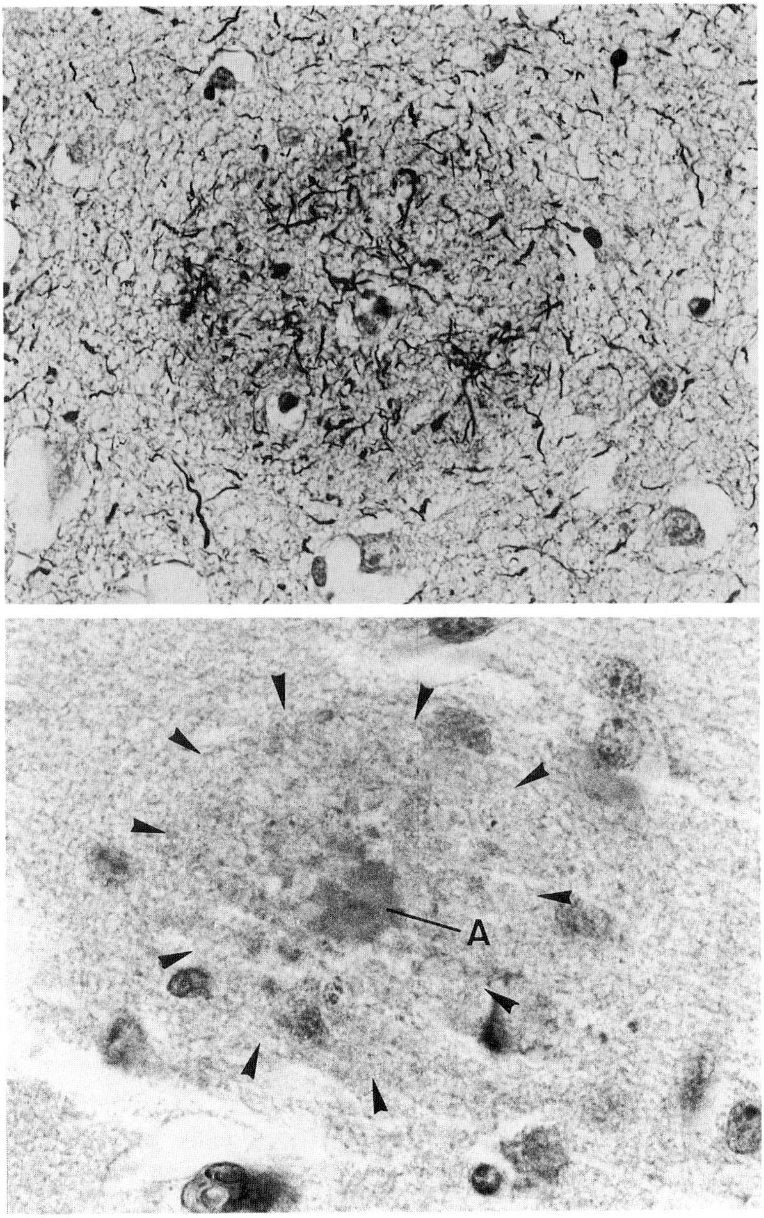

a

b

impregnation of the neocortex in a SDAT patient of group II. Note total absence of threads in the neuropil and absence of impregnated neurites in a plaque of the classical type. A = Amyloid. Plaque is outlined by arrowheads. × 880.

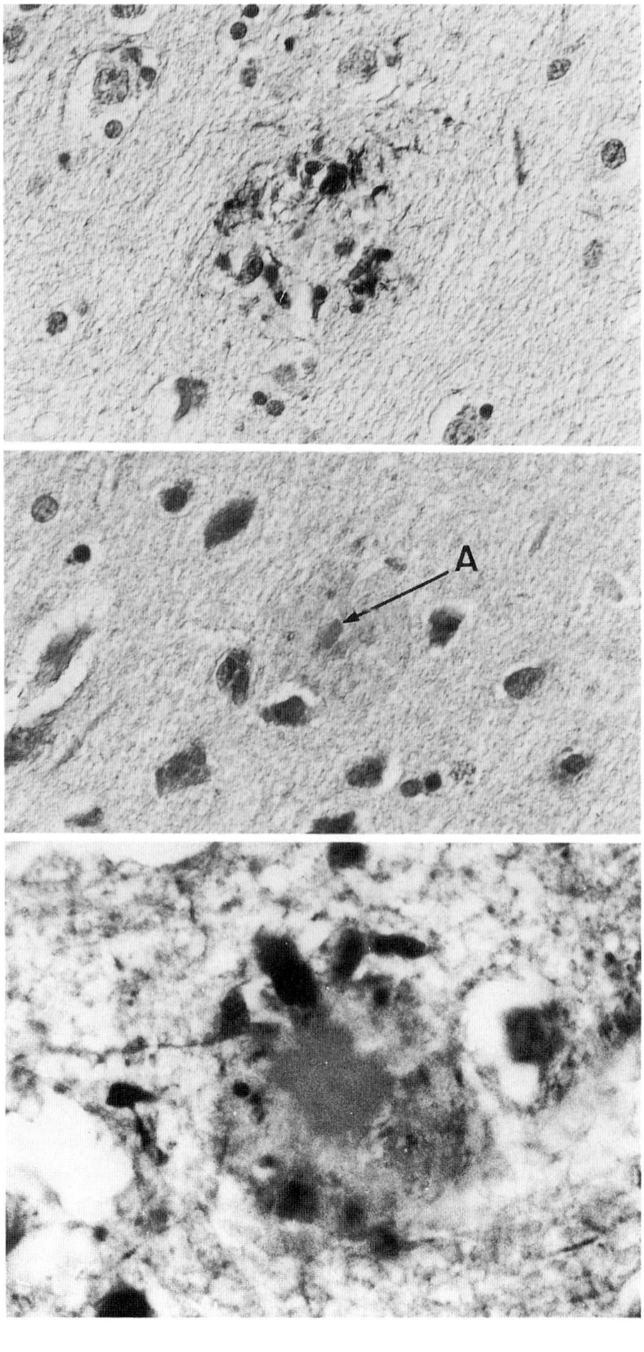

a

b

c

Contrasting with the findings in group I patients, there was a striking absence of Gallyas (fig. 4b) anti-PHF (fig. 5b) and anti-MAP-tau (not shown) labeled neurites in SP of the group II SDAT patients. Absence of labelled neurites was found in all SP irrespective of their maturation stage or morphological type. Total absence of labelled neurites was confirmed by serial sectioning and reconstruction of several plaques through the series. As in group I, SP neurites were consistently labelled with the two monoclonal antibodies RT 97 and BF 10, recognizing phosphorylated neurofilament epitopes (table II, fig. 5c). The number of RT 97 and BF 10 stained neurites in plaques of group II patients and the staining intensity of neurites was comparable to those of the plaques of group I patients.

Control Patients

No threads, NFT or SP could be found in young control patients neither in allocortex nor in neocortex. In old intellectually normal patients, there were occasional Gallyas-positive and immunoreactive NFT and SP, mainly in the entorhinal cortex. Some Gallyas-impregnated threads could also be found but exclusively in the immediate surroundings of NFT. With very few exceptions, the neocortex was free of threads, NFT or SP. BF 10 and RT 97 consistently stained axons in control cases as well as in group I and II patients.

Discussion

The main finding of our investigation consists in a differential pattern of immunoreactivity of SP neurites in the neocortex of SDAT patients depending on the presence or absence of NFT in the same cortical areas.

Fig. 5. a Anti-PHF labeling of a senile plaque in neocortex of a SDAT patient of group I. Note intense labeling of degenerating neurites at the plaque periphery. 4-μm thick paraffin section counterstained with hematoxylin. × 438. *b* Anti-PHF labeling of a senile plaque in neocortex of a group II SDAT patient. No labeled neurites can be found in plaque or in its surrounding. Counterstain with hematoxylin shows the amyloid core of the plaque (A). × 438. *c* Labeling with the monoclonal RT 97 against neurofilament heavy side arm. Neocortex of a group II SDAT patient. Neurites at the periphery of a small plaque are consistently labeled as they do in plaques of group I patients. Amyloid core of the plaque is not stained. 4-μm thick paraffin section, counterstained with hematoxylin. × 880.

There were two main patterns, the one consisting in the presence of NFT, neuropil threads [Braak et al., 1986] and the presence in SP neurites of epitopes and of silver-positive material identical to those of neuropil threads and NFT. The other pattern consisted of SP only, whose neurites were devoid of silver-impregnated and immunoreactive material. Contrasting with these results, however, all SP were found to contain RT 97 and BF 10 phosphorylated neurofilament epitopes irrespective of the presence or absence of NFT.

NFT have been shown to react in situ with a polyclonal anti-PHF serum [Brion et al., 1985a] prepared from SDS-insoluble PHF. Moreover, the serum against isolated NFT has been shown by immunoblotting to contain anti-MAP-tau antibodies [Brion et al., 1985b]. Recent adaptation of the Gallyas method to stain polypeptides separated on SDS polyacrylamide gel has shown that this silver impregnation readily and consistently stains MAP-tau and PHF polypeptide [Iqbal et al., 1987]. The Gallyas method, unlike other silver stains, like for instance Bodian, do not stain normal neurofilaments in axons [Braak et al., 1986; our own observations on control cases]. Furthermore, ultrastructural and immunochemical studies by Braak et al. [1986] have demonstrated that Gallyas-stained neuropil threads contain PHF and that threads are labeled with an anti-PHF antiserum [prepared by Iqbal et al., 1984]. Our observations show that Gallyas-impregnated structures in group I patients are coextensive with MAP-tau and anti-PHF immunoreactivity.

All these findings taken together indicate that with the Gallyas method PHF or/and MAP-tau material is stained selectively. These are also the major antigens in immunocytochemical stains. Our findings confirm results obtained by Braak et al. [1986], demonstrating more extensive cytoarchitectural disruption of the nervous tissue by tau/PHF material than recognized before. However, this is not always the case in SDAT as documented by the present investigation which shows that involvement of the neuropil and especially of SP neurites by MAP-tau and anti-PHF-positive material do not occur in the absence of NFT. Extensive areas of the neocortex in a considerable proportion of SDAT patients do not contain these polypeptides in spite of the presence of numerous SP.

One possible interpretation of these findings is that such polypeptides are not an essential component of the SP and that the development of SP in neocortex is not dependent on the presence of pathological tau and PHF proteins in the neuropil. We may suppose that where NFT develop in neocortical neurons, PHF and tau polypeptide accumulate secondarily in

SP neurites since SP are in part innervated by dendrites and axons of local neurons [Probst et al., 1983]. Our findings also make a local production of PHF and tau polypeptide in the SP unlikely. Furthermore, a formation of amyloid derived from PHF in degenerating neuropil is even more unlikely in view of the present findings. Thus our results are not consistent with the model proposed by Masters et al. [1985]. According to these authors, amyloid fibrils in the SP and in blood vessels are related to intraneuronal PHF and all three fibrous deposits are composed of the same protein. However, as documented immunocytochemically here, a total absence of NFT is compatible with the presence of numerous SP in neocortex. Furthermore, the plaque morphology in cases without tau and PHF-labeled neurites was not different from that of SP bearing labeled neurites. In particular, amyloid-rich plaques were found in similar proportions in the presence or absence of NFT and labeled neurites. Furthermore, as already reported by Selkoe et al. [1986], amyloid of the plaque core did not show labeling using polyclonal and monoclonal antibodies that label PHF (including antibodies to MAP-tau).

In contrast to the differential immunoreactivity with antisera to NFT and MAP-tau, phosphorylated neurofilament epitopes were shared by all neocortical plaques even in the absence of NFT and neuropil threads. This however is not surprising since RT 97 and BF 10 antibodies recognize epitopes of normal neurofilaments in axons. Accumulation of neurofilaments has also been documented in abnormal nerve cell processes of the plaque corona by electron microscopy [Gonatas et al., 1967] and immunocytochemistry using rabbit antineurofilament sera [Anderton et al., 1982; Kahn et al., 1980]. The fact that neurites in SP are stained more consistently than the axons in the surrounding cortex and white matter may be explained by much higher amounts of neurofilament antigens in plaque neurites.

Our findings show that PHF and tau antigens are not an essential or regular component of abnormal neurites found in SP. One may speculate that PHF pathology in neurons is not of pathogenic importance for the formation of SP. Although a primary neuronal dysfunction which escapes our immunochemical investigations cannot absolutely be dismissed as a factor leading to SP formation, the present findings seem at least compatible with a mechanism of SP formation which does not implicate PHF pathology of neurons.

Several hypotheses have been advanced by various authors favoring a nonneuronal origin of SP, especially of its amyloid component [for review, see Wisniewski et al., 1981]. We recently described a type of SP which could correspond to a very primitive stage of SP formation [Probst et al., 1987]. This type of SP regularly contains a central microglial cell and does not contain amyloid deposits at the light microscopic level. Together with the present finding, this observation is compatible with a point source of a neurotoxic factor(s) represented by an activated microglial cell. Spreading around this cell, this factor(s) could induce concentric alterations of the neuropil leading to a spherical lesion, the SP. The same cell might also induce the formation of amyloid fibril protein in the plaque center, for instance by processing a precursor or surrounding normal plaque membranes. Amyloid, however, is only laid down at a time when diffuse alteration of the neuropil around the microglial cell has already occurred indicating that amyloid per se is not the nidus of the plaque.

Conclusions

SP and NFT have long been recognized as the two main histopathological hallmarks of the Alzheimer's disease brain. Despite considerable advances in the molecular pathology of NFT and SP, especially the amyloid component of the plaques, the problem of their origin and relationships remains largely unsolved.

Two main hypotheses have been advanced concerning the relationship of SP and NFT: one suggesting the origin of extracellular plaque amyloid from intraneuronal PHF, the other that PHF originates from vascular or SP amyloid. A third possibility is that PHF pathology in neurons and SP originate independently. To investigate these possibilities we applied a panel of antisera against PHF and MAP-tau polypeptides, a silver impregnation which specifically stains PHF and monoclonal antibodies against epitopes of human neurofilament polypeptide to the study of neocortical SP in two groups of SDAT patients. In the one group, there were numerous NFT as well as SP in the neocortex, in the other group SP without NFT. The main finding of our investigation consists in a differential pattern of immunoreactivity of SP neurites depending on the presence or absence of NFT in the neocortex. In the group with NFT, there were also neuropil threads and SP neurites containing epitopes and silver-positive material identical to those of NFT. In the other group, consisting of SP only, the neurites were devoid of silver-impregnated and immunoreactive material. Contrasting with these results, all SP were found to contain phosphorylated neurofilament epitopes which normally occur in axons irrespective of the presence or absence of NFT. Amyloid in SP remained consistently unstained with the silver method or the antibodies used. Our findings indicate that PHF and tau polypeptide are not essential components of SP, suggesting that the development of SP in the neocortex may occur independently of PHF pathology of neocortical neurons.

References

Anderton, B.H.; Breinburg, D.; Downes, M.J.; Green, P.J.; Tomlinson, B.E.; Ulrich, J.; Wood, J.N.; Kahn, J.: Monoclonal antibodies show that neurofibrillary tangles and neurofilaments share antigenic determinants. Nature, Lond. *298:* 84–86 (1982).

Anderton, B.H.; Haugh, M.C.; Kahn, J.; Miller, C.; Probst, A.; Ulrich, J.: The nature of neurofibrillary tangles; in Traber, Gispen, Senile dementia of the Alzheimer type, pp. 205–216 (Springer, Berlin 1985).

Braak, H.; Braak, E.; Iqbal, I.G.; Iqbal, K.: Occurrence of neuropil thread in the senile human brain and in Alzheimer's disease: a third location of paired helical filaments outside of neurofibrillary tangles and neuritic plaques. Neurosci. Lett. *65:* 351–355 (1986).

Brion, J.P.; Couck, A.M.; Passareiro, E.; Flament-Durand, J.: Neurofibrillary tangles of Alzheimer's disease: an immunohistochemical study. J. submicrosc. Cytol. *17:* 89–96 (1985a).

Brion, J.P.; Passareiro, H.; Nunez, J.; Flament-Durand, J.: Mise en évidence immunologique de la protéine tau au niveau des lésions de dégénérescence neurofibrillaire de la maladie d'Alzheimer. Archs Biol., Brux. *95:* 229–235 (1985b).

Folstein, F.M.; Folstein, S.E.; McHugh, P.R.: Mini-mental test: a practical method for grading the cognitive state of patients for the clinician. J. Psychiat. Res. *12:* 189–198 (1975).

Gallyas, F.: Silver staining of Alzheimer's neurofibrillary changes by means of physical development. Acta morph. hung. *19:* 1–8 (1971).

Gonatas, N.K.; Anderson, W.; Evangelista, I.: The contribution of altered synapses in the senile plaque: an electron microscopic study in Alzheimer's dementia. J. Neuropath. exp. Neurol. *26:* 29–39 (1967).

Iqbal, G.I.; Iqbal, K.; Tung, Y.C.; Wisniewski, H.M.: Alzheimer paired helical filaments: immunochemical identification of polypeptides. Acta neuropath. *62:* 259–267 (1984).

Iqbal, K.; Iqbal, I.G.; Wisniewski, H.: A silver impregnation method specific for Alzheimer paired helical filaments (PHF) stains microtubule associated protein tau. J. Neuropath. exp. Neurol. *46:* 333 (1987).

Kahn, J.; Geon, P.G.; Thorpe, R.; Anderton, B.H.: Immunohistochemistry of neurofilaments in Alzheimer's disease. J. clin. exp. Gerontol. *2:* 199–210 (1980).

Khachaturian, Z.S.: Diagnosis of Alzheimer's disease. Archs Neurol. *42:* 1097–1104 (1985).

Kidd, M.: Paired helical filaments in electron microscopy of Alzheimer's disease. Nature, Lond. *197:* 192–193 (1963).

Kosik, K.S.; Joachim, C.L.; Selkoe, D.J.: The microtubule-associated protein tau is a major antigenic component of paired helical filaments in Alzheimer's disease. Proc. natn. Acad. Sci. USA *83:* 4044–4048 (1986).

Masters, C.L.; Multhaup, G.; Simms, G.; Pottgiesser, J.; Martins, R.N.; Beyreuther, K.: Neuronal origin of cerebral amyloid: Neurofibrillary tangles of Alzheimer's disease contain the same protein as the amyloid of plaque cores and blood vessels. Eur. molec. Biol. Org. J. *4:* 2757–2763 (1985).

Probst, A.; Basler, V.; Bron, B.; Ulrich, J.: Neuritic plaques in senile dementia of Alzheimer type: a Golgi analysis in the hippocampal region. Brain Res. *268:* 249–254 (1983).

Probst, A.; Brunnschweiler, H.; Lautenschlager, C.; Ulrich, J.: A special type of senile plaque, possibly an initial stage. Acta neuropath. *74:* 133–141 (1987).

Schwartz, P.: Amyloidosis: cause and manifestation of senile deterioration (Thomas, Springfield 1970).

Selkoe, D.J.; Abraham, C.R.; Podlisny, B.; Duffy, L.K.: Isolation of low-molecular weight proteins from amyloid plaque fibers in Alzheimer's disease. J. Neurochem. *146:* 1820–1834 (1986).

Terry, R.D.; Hansen, L.A.; De Teresa, R.; Davies, P.; Tobias, H.; Katzman, R.: Senile dementia of the Alzheimer type without neocortical neurofibrillary tangles. J. Neuropath. exp. Neurol. *46:* 262–268 (1987).

Terry, R.D.; Wisniewski, H.M.: The ultrastructure of the neurofibrillary tangle and the senile plaque; in Wolstenholme, O'Connor, Alzheimer's disease and related conditions, pp. 145–168 (Churchill, London 1970).

Ulrich, J.; Stähelin, H.B.: The variable topography of Alzheimer type changes in senile dementia and normal old age. Gerontology *30:* 210–214 (1984).

Wisniewski, H.M.; Sinatra, R.S.; Iqbal, K.; Grundke-Iqbal, I.: Neurofibrillary and synaptic pathology in the aged brain; in Johnson, Aging and cell structure, pp. 165–142 (Plenum Publishing, New York 1981).

Wisniewski, H.M.; Terry, R.D.: Neuropathology of the aged brain; in Terry, Gershon, Neurobiology of aging, pp. 265–280 (Raven Press, New York 1976).

A. Probst, MD, Department of Pathology, Division of Neuropathology, University of Basel, CH-4056 Basel (Switzerland)

Subject Index